- INSTALL ELECTRICAL BREAKERS FOR ENTIRE SHOP WITHIN EASY REACH, CIRCUIT-RATED FOR SUFFICIENT AMPERAGE
- STOCK FIRST AID KIT WITH MATERIALS TO TREAT CUTS, GASHES, SPLINTERS, FOREIGN OBJECTS AND CHEMICALS IN EYES, AND BURNS
- HAVE TELEPHONE IN SHOP TO CALL FOR HELP
- INSTALL FIRE EXTINGUISHER RATED FOR A-, B-, AND C-CLASS FIRES
- WEAR EYE PROTECTION AT ALL TIMES
- LOCK CABINETS AND POWER TOOLS TO PROTECT CHILDREN AND INEXPERIENCED VISITORS
- USE DUST COLLECTOR TO KEEP SHOP DUST AT A MINIMUM
- WEAR SHIRT SLEEVES ABOVE ELBOWS
- WEAR CLOSE-FITTING CLOTHES
- WEAR LONG PANTS
- REMOVE WATCHES, RINGS, OR JEWELRY
- KEEP TABLE AND FENCE SURFACES WAXED AND RUST-FREE
- WEAR THICK-SOLED SHOES, PREFERABLY WITH STEEL TOES

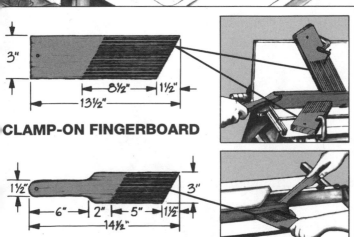

3"

8½" — 1½"

13½"

CLAMP-ON FINGERBOARD

1½"

6" — 2" — 5" — 1½"

3"

14½"

HAND-HELD FINGERBOARD

PROTECTION

WEAR FULL FACE SHIELD DURING LATHE TURNING, ROUTING, AND OTHER OPERATIONS THAT MAY THROW CHIPS

WEAR DUST MASK DURING SANDING AND SAWING

WEAR VAPOR MASK DURING FINISHING

WEAR SAFETY GLASSES OR GOGGLES AT ALL TIMES

WEAR RUBBER GLOVES FOR HANDLING DANGEROUS CHEMICALS

WEAR EAR PROTECTORS DURING ROUTING, PLANING, AND LONG, CONTINUOUS POWER TOOL OPERATION

THE WORKSHOP COMPANION™

USING THE BAND SAW

TECHNIQUES FOR BETTER WOODWORKING

by Nick Engler

Rodale Press
Emmaus, Pennsylvania

Printed in the United States of America on acid-free ∞, recycled ♲ paper

If you have any questions or comments concerning this book, please write:
  Rodale Press
  Book Readers' Service
  33 East Minor Street
  Emmaus, PA 18098

About the Author: Nick Engler is an experienced woodworker, writer, and teacher. He worked as a luthier for many years, making traditional American musical instruments before he founded *Hands On!* magazine. Today, he contributes to several woodworking magazines and teaches woodworking at the University of Cincinnati. This is his thirtieth book.

Series Editor: Jeff Day
Editors: Roger Yepsen
        Bob Moran
Copy Editor: Sarah Dunn
Graphic Designer: Linda Watts
Graphic Artists: Mary Jane Favorite
        Chris Walendzak
Photographer: Karen Callahan
Cover Photographer: Mitch Mandel
Proofreader: Hue Park
Indexer: Beverly Bremer
Typesetting by Computer Typography, Huber Heights, Ohio
Interior and endpaper illustrations by Mary Jane Favorite
Produced by Bookworks, Inc., West Milton, Ohio

Library of Congress Cataloging-in-Publication Data

Engler, Nick.
    Using the band saw/by Nick Engler.
        p.   cm. — (The workshop companion)
    Includes index.
    ISBN 0–87596–140–1   hardcover
    1. Band saws. 2. Woodwork. I. Title II. Series:
Engler, Nick.   Workshop companion.
TT186.E543   1992
684'.083—dc20   92–25296
                                    CIP

2 4 6 8 10 9 7 5 3   hardcover

Special Thanks to:

Jim Berkley
Shopsmith, Inc.
Vandalia, Ohio

Delta International Machinery
  Corporation
Pittsburgh, Pennsylvania

Grizzly Imports, Inc.
Bellingham, Washington

Chuck Olson
Olson Blade Company
Medford, Connecticut

CONTENTS

TECHNIQUES

PROJECTS

TECHNIQUES

1

CHOOSING A BAND SAW

There are few power tools as versatile as the band saw. You can use this tool to make crosscuts, rips, miters, and bevels. It will cut in a straight line as well as make curves and contours. It will saw both thin and thick stock with equal ease. And it will cut a variety of materials — wood, wood products, plastic, and metal. Furthermore, there are several woodworking operations that you can't perform on any other power tool, such as cutting cabriole legs and resawing wide stock. It's no wonder that the band saw is indispensable to many experienced woodworkers.

The band saw is the result of several nineteenth-century technological innovations. Saws with thin, narrow blades, such as the bow saw and the buck saw, have been used since early Roman times. In 1808, Englishman William Newbury conceived the idea of welding the ends of a long buck saw blade together to make a continuous loop, then mounting this loop on two rotating wheels. One wheel was positioned above a cutting table, and the other below it.

Unfortunately, Newbury's idea was ahead of his time. In the early 1800s, buck saw blades were too brittle and the welds too fragile to make the band saw a practical tool. But in 1846, a French woman, Mlle. Crepin, developed a reliable method of joining the ends of blades by *brazing* them. Around the same time, metallurgists found

a way to roll flexible "spring" steel into thin blades. By the second half of the nineteenth century, band saws were used throughout the woodworking industry. And by the early twentieth century, tool manufacturers offered reasonably priced models that were small enough to be used in modest-size businesses and home workshops.

BAND SAW FEATURES

MAJOR PARTS

All band saws, large and small, have several important parts in common (*SEE FIGURE 1-1*):

■ The *blade* is a continuous band of flexible steel with teeth ground in one edge.

■ This blade revolves on two or more *wheels*. The wheel that drives the blade is called the *drive wheel*, and the others are *idler wheels*. On most two-wheel band saws, the drive wheel is on the bottom and the idler wheel is on the top.

■ The wheels are covered with hard rubber *tires* to cushion the blade and protect the teeth.

■ The *motor* turns the drive wheel, which drives the blade. On some band saws, the motor is coupled to the drive wheel by pulleys and a V-belt. On others, the drive wheel is mounted directly on the motor shaft.

■ At least one wheel is mounted above a *worktable* and another below it. As the blade revolves on the wheels, it goes through this table. The cutting action of the blade holds the work down on the table.

■ There are two guides — an *upper blade guide* and a *lower blade guide* — to keep the blade running straight and true where it goes through the table.

■ The wheels, worktable, blade guides, and, on some smaller band saws, the motor, are mounted on a *frame*. This frame is usually a casting made from metal or plastic.

■ The frame is divided into two parts, upper and lower, with a narrow *column* in between.

■ The horizontal distance between the column and the blade is called the *throat*. This determines one capacity of the band saw — how far it can cut in toward the center of a workpiece.

■ The frame and column are fitted with a plastic or metal *cover* that helps protect you from the unused portion of the band saw blade as it revolves on the wheels.

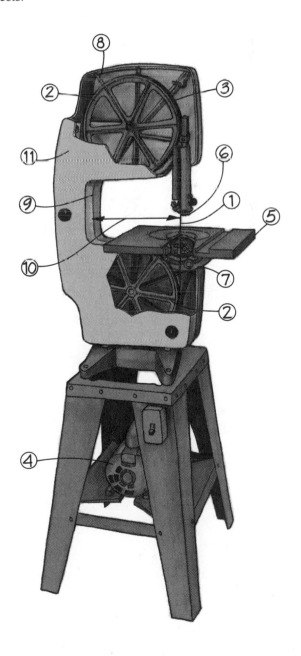

1-1 The most important part of any band saw is the *blade* (1), a long, thin, continuous loop mounted between two or more *wheels* (2) with hard rubber *tires* (3). A lower wheel is driven by a *motor* (4), which, in turn, drives the blade. The blade goes though a *worktable* (5) and is prevented from flexing or drifting by an *upper blade guide* (6) above the worktable and a *lower blade guide* (7) below it. The wheels, worktable, and guides are mounted on a casting or *frame* (8). This frame consists of two sections, upper and lower, separated by a *column* (9). The horizontal distance between the column and the blade is the *throat* (10). The frame and the column are protected by a *cover* (11).

IMPORTANT SUBASSEMBLIES

Several of these major features — the worktable, blade guides, and idler wheels — are further divided into important subassemblies.

The parts of the *worktable* hold the work as it's cut (SEE FIGURE 1-2):

■ Where the blade goes through the table, a removable *table insert* reduces the opening and provides additional support for the work. It's usually made of a soft metal, so as not to damage the teeth should the blade drift into the insert.

■ A *blade mounting slot* in the table allows you to remove and replace the band saw blade.

■ A *leveling pin* joins the edges of the table together on either side of the slot and keeps the table flat. On some inexpensive models, the leveling pin is replaced by a simple *table tie*.

■ The table is mounted on *trunnions* so it can be tilted.

■ A *trunnion lock* secures the trunnions, holding the table square or at some angle to the blade.

■ Newer band saws have a *miter gauge slot* to guide a miter gauge.

The blade guides support the blade on three sides (SEE FIGURE 1-3):

■ Both the upper and the lower blade guides hold *guide blocks* that keep the blade from twisting as it cuts. These blocks are mounted on either side of the blade and can be moved closer to or farther away from the blade, as needed.

■ *Thrust bearings* or *thrust plates* on both guides prevent the blade from being deflected backward. These, too, can be adjusted closer to or further from the blade.

■ The upper blade guide is mounted on a *guide post*. The guide and post can be raised or lowered to accommodate both thick and thin stock.

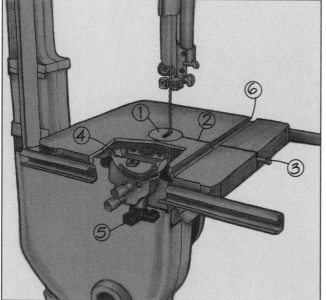

1-2 The band saw worktable has a soft metal *blade insert* (1) to protect the blade teeth and provide additional support for the work around the blade. A *blade mounting slot* (2) lets you change blades easily, and a *leveling pin* (3) prevents the table from twisting or warping at the slot. The worktable tilts on *trunnions* (4), and these trunnions are secured by a *trunnion lock* (5). The worktable may have a *miter gauge slot* (6) running parallel to the face of the blade.

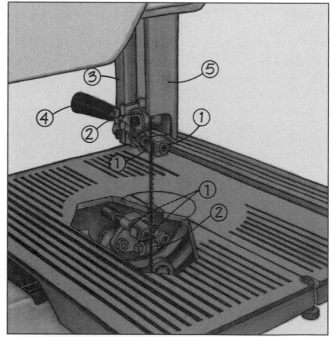

1-3 The blade guides hold *guide* *blocks* (1) to keep the band saw blade from twisting, and *thrust bearings* (2) to keep the blade from flexing backward. The upper blade guide is mounted on a movable *guide post* (3), and can be secured at any height by a *guide lock* (4). The unused portion of the blade between the frame and the upper blade guide is covered by a *blade guard* (5).

■ A *guide lock* secures the upper blade guide and the guide post at the desired height.

■ A *blade guard,* mounted on the guide post, protects you from the unused portion of the blade between the upper frame and the upper blade guide.

On two-wheel band saws, the idler wheel can be moved up or down to control blade tension. On *most* two-wheel saws, this idler wheel can also be tilted from side to side to adjust the blade tracking. (A few two-wheel band saws, such as the Shopsmith, have automatic tracking devices and neither wheel tilts.) On three-wheel band saws, the tensioning and tracking adjustments may be incorporated in one idler wheel, or they may be assigned to separate wheels (*SEE FIGURE 1-4*):

■ A *tension adjustment* moves the idler wheel up or down to keep the blade under tension. The blade must be tensioned to keep it straight in the cut, and the wider the blade, the more tension it requires. **Note:** On some three-wheel band saws, the tension adjustment is on one of the bottom wheels.

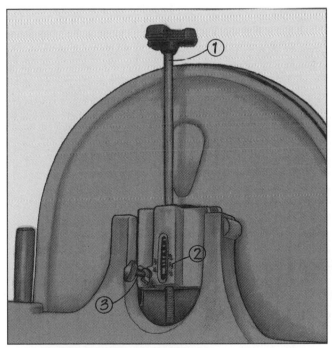

1-4 The top band saw wheel can be moved up or down by the *tension adjustment* (1) to change the tension on the blade. A *tension scale* (2) shows how tight or loose the blade is. The top wheel can also be tilted from side to side by a *tracking adjustment* (3) to position the blade on the wheels.

■ A *tension scale* shows how much tension is being applied to the blade. This scale is usually calibrated for blade width (1/8 inch, 1/4 inch, and so on).

■ A *tracking adjustment* tilts the top wheel from side to side. This moves the running blade toward the high portion of the wheel and lets you position the blade front to back on the band saw.

TYPES OF BAND SAWS
TWO-WHEEL AND THREE-WHEEL SAWS

The many band saws designed for small workshops can be grouped into two categories — *two-wheel* and *three-wheel* band saws. There have also been a few four-wheel band saws, but they didn't offer significant advantages and were discontinued.

Two-wheel band saws require large wheels for a useful throat capacity. And the larger the wheels, the wider the blade you can mount on them. (Wide blades are not as flexible as narrow ones, and cannot bend around small wheels.) Because large blades require more blade tension, the band saw frame must be substantial enough to withstand the load. The big wheels and the heavy frame make these tools more massive than their three-wheel cousins. Consequently, two-wheel band saws are usually large, stationary power tools. (*SEE FIGURE 1-5.*)

The third wheel of a three-wheel saw allows a wider throat by routing the blade further from the guides. This allows you to cut wide workpieces, although you can only use narrow blades. Since narrow blades don't require as much tension as larger ones, the frame can be much lighter. Many three-wheel band saws are small and light enough to be portable benchtop tools, but have sufficient capacity to cut large workpieces. (*SEE FIGURE 1-6.*)

Both designs offer important advantages. Because the wheels of a two-wheel saw are more massive, they act as flywheels — their inertia keeps the saw running at cutting speed through heavy work. The machine won't bog down as easily, and the resulting cut is smoother and more consistent. Also, the blade flexes a great deal less as it goes around two large-diameter wheels than it does on three smaller wheels. The thin metal doesn't fatigue as quickly, and the blade lasts longer. On the other hand, many three-wheel saws offer the same (or larger!) throat capacity as two-wheel tools costing hundreds of dollars more. They often require less space and less power, and can be stored under a workbench.

1-6 Three-wheel band saws are usually (but not always) smaller and more portable than the two-wheel variety. While these saws are best suited for narrow blades and light work, the wheel arrangement offers a large throat capacity so you can cut wide stock.

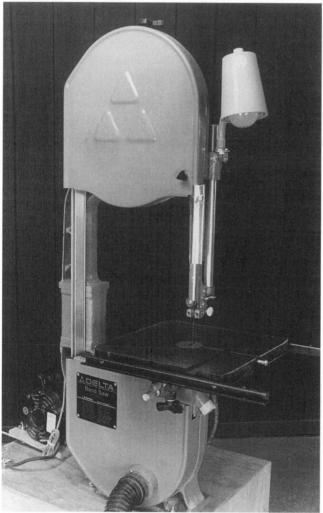

1-5 Two-wheel band saws usually have large wheels and heavy frames. This makes them larger than three-wheel models, but they are better suited for heavy work. Two-wheel band saws, by virtue of their large-diameter wheels, will accept wider blades than three-wheel saws.

1-7 A *horizontal* band saw works like a miter saw or "chop" saw — the tool moves while the work remains stationary. It's designed to cut metal pipes and bar stock to length, but many horizontal saws can be locked in a "vertical" mode and used to cut wood.

Two common variations on the two-wheel band saw are useful in small shops. A *horizontal band saw* is designed to cut off lengths of metal, but will do double duty as a wood-cutting band saw. (*SEE FIGURE 1-7.*) If you work with both wood and metal, consider one of these. On a *tilting-frame band saw,* the frame tilts while the table remains horizontal. (*SEE FIGURE 1-8.*) This lets you cut angles while keeping the work flat.

1-8 Tilting-frame band saws were originally developed for the boat-building industry to make beveled cuts in long workpieces, and they are now available for home work-shops. Because the work remains horizontal while you tilt the saw, it's much easier to cut angles.

WHAT TO LOOK FOR

When choosing a band saw for your workshop, con-sider these features:

Cutting capacity — The throat on a band saw deter-mines how *wide* a board it will cut; the maximum depth of cut — the distance from the worktable to the upper blade guide when the guide is raised as high as it will go — determines how *thick* the work-piece can be. (*SEE FIGURE 1-9.*) If you work with large, broad sheets, you may need a band saw with a large throat. However, don't be overly concerned with throat capacity — you can cut wide workpieces on saws with narrow throats simply by planning your cuts carefully. It's much more important to have the depth of cut you need. Most home workshop band saws will cut a workpiece between 4 and 6 inches thick, but if you work with thicker stock or do a lot of resawing (cutting wide boards on edge), you may require more depth capacity.

Blade capacity — Narrow blades cut intricate curves and contours; wide blades are best used for resawing and cutting thick stock. (For more information on blade use, refer to "Band Saw Blades" on page 14.) The use of a band saw is limited by its blade capacity — the widest blade it will mount. Most three-wheel benchtop band saws will only accept blades up to $\frac{1}{4}$ or $\frac{3}{8}$ inch wide and are best used for scrollwork.

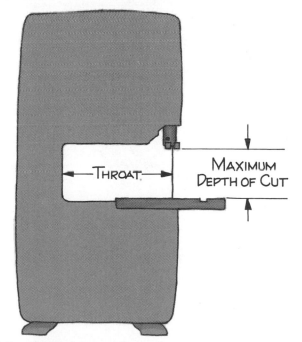

1-9 The thickest workpiece you can cut (or the widest workpiece you can resaw) is determined by the band saw's *maximum depth of cut.* To find this capacity, raise the upper blade guide as high as it will go and measure the distance from the guide to the table.

Medium-size two-wheel saws usually drive blades up to ½ inch wide, and larger home workshop saws can use ¾-inch-wide blades — sufficient for occasional resawing jobs. Only the largest saws can mount blades 1 inch wide or wider. These can be used continuously to cut extremely thick stock.

Wheels and tires — These are the most important components of the band saw. For the blade to run smoothly, the wheels must be perfectly round. If one or more wheels are out of round, even slightly, the tool will vibrate as it runs, and the blade won't cut evenly. *(SEE FIGURE 1-10.)* Hold a scrap of wood against the band saw frame so one face is about ¹⁄₃₂ inch away from a tire. Slowly rotate the wheel — if the surface of the tire moves closer to or farther away from the wood block, the wheel is out of round. **Note:** The prevailing myth is that metal wheels are better than

plastic, but this is not necessarily so. Recent advances in plastics technology allow manufacturers to make plastic wheels that are just as durable and precise as the best metal wheels!

The wheels should be covered with hard rubber tires, glued onto the rim. On some band saws, such as Delta, all tires are crowned — this makes it easier to track the blades. On many other band saws, such as Inca and Shopsmith, all tires are flat — this provides better blade support, especially for wide blades. On a few band saws, the idler tire with the tracking adjustment is crowned, while the other tires are flat. This arrangement is a good compromise between the two.

Frame — The frame should be as sturdy as possible, particularly in the area of the column. If the frame is weak, it will distort as you apply tension to the blade, making it difficult to align and adjust the

FOR YOUR INFORMATION

Out-of-round pulleys and poorly made V-belts can cause vibration, which is transferred along the drive train to the blade. Although this isn't as serious as vibration caused by out-of-round wheels, it may still interfere with the quality of the cut. To eliminate this vibration, replace stamped and die-cast pulleys with machined *cast-iron* pulleys, and ordinary V-belts with *link-type* belts, as shown. Both cast-iron pulleys and link-type belts are available from sources that sell electric motors.

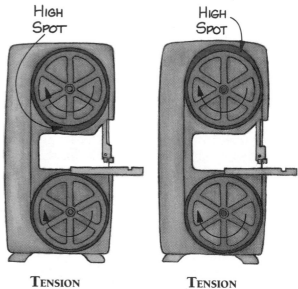

HIGH SPOT   HIGH SPOT

TENSION RELAXED   **TENSION INCREASED**

1-10 Out-of-round wheels may cause the band saw to cut poorly. On these drawings of a two-wheel band saw, one wheel has an exaggerated high spot, making it slightly egg-shaped. When the blade is running over the low portion of the wheel, the tension is relaxed. When it runs over the high spot, the tension increases. As the wheels revolve, the tension constantly fluctuates. This sets up a vibration in the blade and makes the cut uneven. It will also cause the blade to break prematurely.

band saw accurately. The larger the blade used, the more tension you must apply, and the stronger the frame must be. Cast-iron frames are the strongest, followed by cast aluminum. Plastic, although it makes good band saw wheels, is not well-suited for frames. Most plastic materials "creep" (bend very slowly) when under tension, and may become permanently distorted after a time.

Blade guides, tensioning, and tracking — The blade tension, tracking, and guides must be readjusted every time you change the band saw blade. Consequently, you want the adjustment knobs and set screws to be easily accessible. Also look for *precision*. If the blade guides are poorly machined or the blocks fit sloppily, they will be difficult to adjust even if you can reach the set screws easily. Some blade guides — particularly those on horizontal band saws — let you twist the blade at 45 degrees to the normal plane. (*SEE FIGURE 1-11.*) This is a handy feature if you need to do a lot of cut-off work.

Note: If you choose a band saw with an automatic tracking device, make sure that device allows for *some* adjustment so you can fine-tune the tracking if needed.

Worktable — The worktable on most band saws is small compared to other stationary and benchtop power tools. You don't want it too big; otherwise, you couldn't reach the blade comfortably. However, it must be large enough to safely support the workpiece. In some cases, you may need to purchase or build a table extension. (Refer to *Auxiliary Band Saw Tables* on page 47 for more information.) More important than the size is the *flatness*. Band saw tables must have a slot cut in them so you can change blades, and they tend to warp or twist at this point. One side of the slot may become higher than the other, creating a step. This makes it difficult to feed a workpiece, and impossible to cut it accurately. To prevent table warp, look for tables with a *leveling pin* rather than a simple table tie. (*SEE FIGURE 1-12.*)

You should also consider the worktable tilt. Most band saw tables tilt 45 degrees *away* from the column, but only a few degrees *toward* it before the table is stopped by the lower frame. If you look around, you can find band saws that tilt 5 or 10 degrees toward the column. This extra tilt capacity is handy for a few specific woodworking operations, but not absolutely necessary. You can work around it by making a tilted jig for the operation you wish to perform.

Motor, power, and speed — If you will be working for long periods of time at a band saw, look for a model with a heavy-duty capacitor-start or *induction* motor. *Universal* motors, such as those found in portable and some benchtop power tools, are suitable only for short operations. They will not stand up to

1-11 On some band saws, you can reconfigure the blade guides to twist the blade at 45 degrees to its normal plane. This makes it possible to cut long stock to length without the interference of the column, regardless of throat capacity. However, twisting the blade causes additional metal fatigue and friction, and the blade breaks much sooner than with normal use.

1-12 A leveling pin helps prevent the worktable from twisting at the slot. Simply tap it *gently* in place with a hammer after changing the blade. To remove it, twist it with a wrench.

continuous use. The power you need depends on what sort of work you will be doing. A motor that generates ½ horsepower (under load) will be sufficient for most light band saw work. If you do occasional resawing of boards up to 6 inches wide, ¾ horsepower will serve you better. To resaw stock up to 12 inches wide, purchase a motor rated for at least 1½ horsepower.

The motor should drive the blade at the proper speed for the materials you're cutting. Most band saws are set up to run between 2,600 and 3,100 fpm (feet per minute). This is a suitable speed for general cuts in wood. However, when ripping thin stock or cutting sharp curves, a speed between 1,000 and 2,000 fpm will work better. (Be careful not to run the blade too slowly when cutting wood; it won't clear the sawdust from the cut properly.) Make cuts in plastic between 700 and 1,000 fpm, and in metals between 40 and 400 fpm.

Note: Only a few band saws have adjustable speeds. However, you can easily make your own speed changer for any belt-driven band saw by adding two step pulleys and an intermediate arbor between the motor and the drive wheel. (SEE FIGURE 1-13.)

1-13 If you need to run your band saw at several different speeds, make a speed changer from two step pulleys and an extra arbor. For complete instructions, refer to "Band Saw Speed Changer" on page 20.

FOR YOUR INFORMATION

To calculate the speed of a band saw blade in feet per minute (fpm), you must know the revolutions per minute (rpm) of the motor, the diameter of the drive wheel, and the diameter of the pulleys (if any). To find the speed of a belt-driven band saw, use this equation:

Blade speed in fpm
= motor rpm x [(motor pulley diameter - ¼)
÷ (drive pulley diameter - ¼)]
x [(drive wheel diameter x 3.1416) ÷ 12]

For instance, if a band saw with a 12-inch-diameter drive wheel is powered by a motor running at 1,725 rpm, the motor pulley is 2½ inches in diameter, and the drive pulley is 4½, then the blade will run at 2,869 fpm:

1,725 x [(2½ - ¼) ÷ (4½ - ¼)]
x [(12 x 3.1416) ÷ 12] = 2,869

Note: When figuring the speeds of belt-driven machines, you must subtract ¼ inch from each pulley diameter. This is because the belt rides in a groove and contacts the pulley at a spot just inside the rim. So the working diameter (commonly called the *pitch diameter*) of a pulley is slightly smaller than the outside diameter.

To find the speed of a direct-drive band saw, use this equation instead:

Blade speed in fpm
= motor rpm x [(drive wheel diameter
x 3.1416) ÷ 12]

A band saw with a 6-inch-diameter drive wheel powered by a 1,725-rpm motor would run at 2,710 fpm:

1,725 x [(6 x 3.1416) ÷ 12] = 2,710

So — when all is said and done, what sort of band saw should you choose? This depends not only on what sort of work you want to do but also on the money you have to spend and the space you have in your shop. If you need a general-purpose band saw and have the money and the space, a medium-size or large model with cast metal wheels and frame will do well. However, if you have space restrictions or only require a light-duty tool, a smaller, less expensive saw will also serve. If you have never owned a band saw before, you'll quickly find that because of its versatility, *any* well-made model is worth its weight in gold.

2

BAND SAW BLADES AND ACCESSORIES

There are many different band saw accessories to help extend the capabilites of the tool. You can purchase a miter gauge or a fence to help guide the stock as you cut it, or a table extension to help support the stock. A riser block allows you to cut thicker stock, and cool blocks extend the life of your blade. There are even kits to convert your band saw to a strip sander.

The most useful accessories, however, are band saw blades. There are dozens of blades available in different widths and with different tooth patterns. Each blade is designed to perform a specific task, such as cutting intricate patterns or resawing thick stock. With a good selection of blades for your saw, you can make almost any cut in any material.

BAND SAW ACCESSORIES

AVAILABLE ATTACHMENTS

Although the band saw is not as heavily "accessorized" as other power tools, there are several attachments that are commonly available:

Miter gauge — This is one of the least useful accessories. You cannot easily align the miter gauge slot with the blade and even if you could the blade might not cut parallel to the miter gauge slot. (For more information on this subject, refer to "Making Guided Cuts" on page 44.) For this reason, miter gauges can only be used when crosscutting very narrow stock, such as dowels. Older band saws didn't have miter gauge slots; manufacturers began to offer them on newer models as a marketing gimmick. If your band saw doesn't come with a miter gauge, don't bother

purchasing one; you won't miss it. In fact, with a fence and a rectangular wood scrap, you can perform most miter gauge operations. (See *TRY THIS TRICK* on page 45.)

Fence — While fences are useful in ripping and resawing operations, commercially made fences are worthwhile *only* if you can adjust the fence angle in relation to the blade. (*SEE FIGURE 2-1.*) Band saw blades often don't cut parallel to the blade plane; they usually *lead* to one side or another. Whenever you change blades, you must change the fence angle to compensate for this. (Again, refer to "Making Guided Cuts" on page 44.) For this reason, it may be better to make your own adjustable band saw fence.

2-1 For a band saw fence to be useful, you must be able to adjust the fence angle to compensate for blade lead. The fence shown can be adjusted by loosening two bolts on the top of the fence, setting the fence to the desired angle, then tightening the bolts again.

2-2 A *riser block* increases the length of the column, which, in turn, increases the maximum depth of cut. This is an inexpensive way to get a band saw with the capacity to cut extremely thick workpieces.

Riser blocks — Some band saws have two-piece frames, split at the column. You can purchase spacers to insert between the upper and lower portions of the frame and increase the maximum depth of cut. This lets you cut thicker workpieces and resaw wider boards. (*SEE FIGURE 2-2.*) If you purchase a riser block kit for your saw, make sure you also get a longer guide post and blade guard for the upper blade guide.

"Cool" guide blocks — Most guide blocks are made from metal. As the blade slides between the blocks, the friction heats the blade. If the blade is wide enough, this heat quickly dissipates. But on narrow blades, the heat builds up and may cause the blade to break prematurely. "Cool" blocks are made from graphite-impregnated resin, which greatly reduces the friction, prevents heat build-up, and extends the blade life. (*SEE FIGURE 2-3.*) However, because the resin is soft in

comparison to metal, the blocks must be resurfaced frequently.

Table extensions — A few manufacturers offer an extension for band saw tables, similar to the "wings" that come with most table saws. While these do provide additional support, they only extend the worktable in one direction. You can easily make your own extension to extend the table in three directions. (Refer to *Auxiliary Band Saw Tables* on page 47 for instructions.)

Strip sanding kits — You can convert your band saw into a giant strip sander with one of these kits. Mount a ½-inch-wide sanding belt, the same circumference as the band saw blade, on the wheels, then attach a platen to the table to back up the sanding belt. (*SEE FIGURE 2-4.*) **Note:** This conversion is time-consuming and the abrasive belts may damage the band saw if mounted incorrectly.

2-3 "Cool" blocks greatly reduce blade friction, extending the blade's life. Furthermore, because these blocks are softer than metal blades, you can use them to completely surround a blade without damaging the teeth. This provides additional support for a narrow blade when doing scrollwork.

2-4 A strip sanding kit turns your band saw into a giant strip sander. However, to use this sanding accessory without damaging the band saw, you must make sure that the abrasive belt does not contact any portion of the upper and lower blade guides. On some machines, you must remove the guides completely.

BAND SAW BLADES

BLADE FEATURES

It's not enough to choose a good band saw. You must also have a selection of blades and be able to choose the right blade for the job. To do this, it helps to understand the various components of a blade (*SEE FIGURE 2-5*):

■ The *body* is the flexible steel band from which the blade is made.

■ The *gauge* is the thickness of the body — the smaller the gauge, the more flexible the body. Most blades for home workshop use have a gauge between .014 and .025 inch.

■ The *weld* is where the body is joined to make a loop. On many small-gauge blades, this isn't really a weld, but a silver brazing.

■ The *teeth* are ground into the front edge of the body.

■ The *back* edge, where the body rides against the thrust bearing, remains smooth.

■ Each tooth has a cutting *point* and a cutting *face*.

■ The face of each tooth is ground at a particular *rake angle* to a line drawn perpendicular to the back. The rake angle on most band saw blades is between 0 and 10 degrees.

■ The *gullets* are spaces between the teeth that clear the sawdust from the cut. The larger the gullet, the more sawdust it will hold and the faster the blade will cut.

■ The *width* of the blade is the distance between the tips of the teeth and the edge of the back.

■ The teeth are bent or *set* slightly to one side or the other, so they will cut a *kerf* slightly wider than the blade is thick. This set is measured from the tip of the tooth bent farthest to the right to that bent farthest to the left. The blade width and the kerf together determine the diameter of the smallest curve the blade will cut.

2-5 Every band saw blade is made from a flexible band of metal, called the *body* (1). The *gauge* (2) is the thickness of the body — the smaller the gauge, the thinner and more flexible the body. The body is joined end to end in a continuous loop by a *weld* (3). The *teeth* (4) are ground into one edge of the body, while the other edge, or *back* (5), remains smooth. Each tooth has a *point* (6) and a *face* (7) that do the cutting. This face is ground at a particular *rake angle* (8) to the back. The spaces between the teeth are called *gullets* (9), and the distance between the tips of the teeth and the back is the *width* (10). The *tooth spacing* (11) is the distance between teeth, and the *pitch* (12) is the number of teeth per inch. The teeth are usually bent or *set* (13) to one side or the other, so the *kerf* (14) left by the blade will be wider than the gauge. The difference between the kerf and the gauge is the *side clearance* (15). This clearance makes it possible to cut curves with the band saw blade.

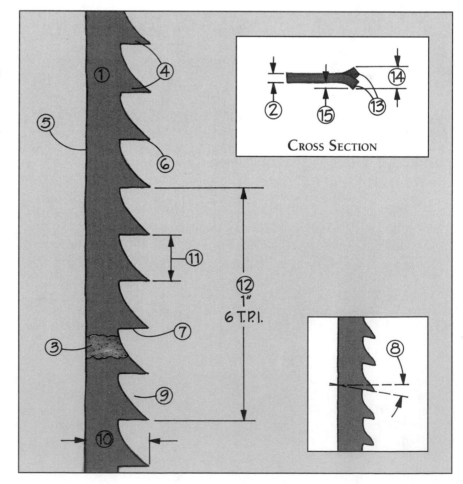

■ The *tooth spacing* is the distance between teeth.

■ The *pitch* is the number of teeth along a length of the blade, normally measured in teeth per inch (TPI).

■ The *side clearance* is the gap between the body of the blade and the side of the kerf. This clearance allows the blade to turn slightly in the kerf, so you can cut curves and contours. (*SEE FIGURE 2-6.*)

CHOOSING A BLADE

When choosing a blade, first decide on its *width*. This, more than any other feature, defines the purpose of the blade. Next, determine the shape or *grind* of the teeth — the configuration of the tooth, gullet, point, face, and rake angle. Finally, consider the *pitch* or spacing of the teeth on the blade.

Width — The width of the blade you need is determined by the intricacy of the pattern you wish to cut. The smaller the diameter of the curves in the pattern, the narrower the blade you need. (*SEE FIGURE 2-7.*) If you want to make gentle curves or straight cuts, you can use a wide blade.

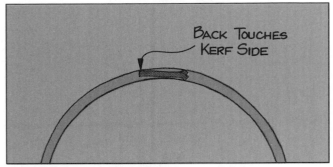

2-6 **Because there is a small** clearance between the side of the kerf and the blade, the blade will turn slightly in the kerf. This enables you to cut curves and contours. The blade stops turning in the kerf when the back hits the side of the kerf. This is why you can't cut small curves with large blades.

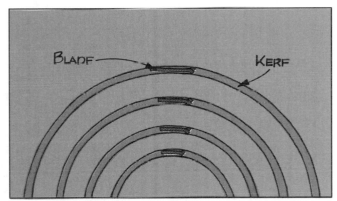

2-7 **When cutting curves, the** smallest radius you can cut with a given blade depends on when the back of the blade hits the sides of the kerf. The backs of wide blades rub against the wood much sooner than narrow blades. For this reason, you can't cut as small a radius with a wide blade as you can a narrow one. Note: If you attempt to cut a curve that's too small for a blade, the blade will twist and the wood will burn.

TRY THIS TRICK

To help choose the blade width, compare the radii in the pattern with a round object whose size corresponds to the smallest radius that can be cut by a specific blade. For example, a ³/₁₆-inch blade will cut a circle roughly the size of a dime. Place a dime on the layout and compare its curved edge to the curves in the pattern. If you find any curves that are smaller, try a pencil eraser. This is about the radius of the smallest curve you can cut with a ¹/₈-inch blade. You may wish to cut a set of different-size circles from paper or cardboard to use when making these comparisons.

The standard woodworking wisdom is to choose the width of the blade for the smallest curves in the pattern. However, this criteria won't always get you the best blade for the job. If you're cutting thick stock or doing continuous band saw work and you have a few tight curves to cut, you don't necessarily want work with a narrow blade. In these cases, you may do better with a wider blade for most of the operation. When it's time to cut the smaller curves, you can either change to a narrower blade or use one of the techniques described in "Making Freehand Cuts" on page 41 for cutting small curves with wide blades.

FOR BEST RESULTS

If the band saw operation will generate a lot of friction, such as cutting green wood or metal, choose a wide blade. The wider the blade, the better it will endure the heat caused by friction.

Grind — There are three common shapes of band saw teeth — *standard, skip,* and *hook. (SEE FIGURE 2-8.)* Select the proper grind according to how you will be cutting the wood:

■ For crosscutting (cutting *across* the grain), either a standard or skip grind works well. Standard teeth leave a smoother surface, but a skip-tooth blade cuts faster.

■ When ripping or resawing (cutting *with* the grain), a hook grind works best. Standard and skip teeth tend to follow the grain in a rip cut and may drift to one side or the other if the grain is not perfectly parallel to the direction of the cut. Hook teeth, because of their rake angle, do not follow the grain.

■ For cutting curves and contours (cutting in *all* directions), a skip grind cuts with the least effort. However, if a smooth surface is important, use a blade with standard teeth.

There are also some specific woodworking operations that require particular blade grinds. For example, when

cutting thick stock, use skip or hook teeth — the larger gullets help to clear the sawdust from the cut. When cutting green wood or resinous wood (such as pine, cherry, or teak), use hook teeth. Here again the gullet size is important — the wet or gummy sawdust will pack small gullets, loading the blade and interfering with the cutting action. Since hook teeth have the largest gullets of the three common grinds, the chance of loading is reduced. When making joinery cuts, a smooth surface is important; use standard teeth.

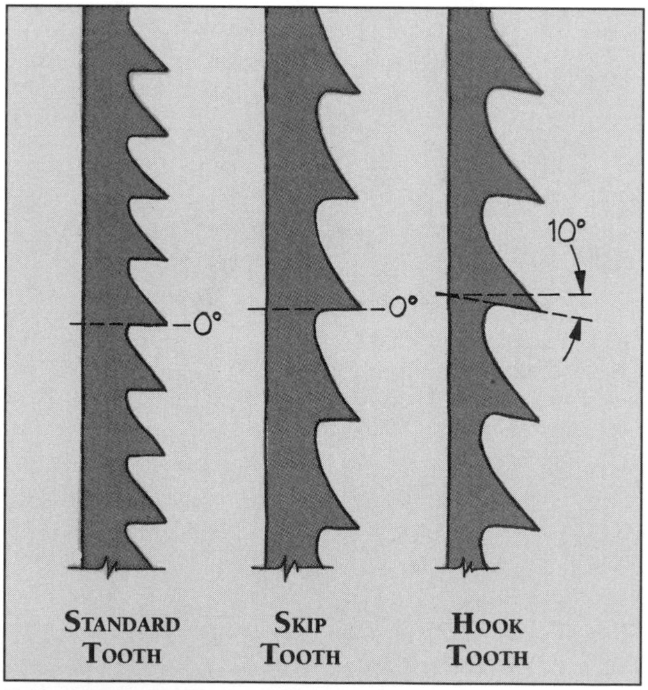

FOR BEST RESULTS

Standard-grind teeth actually cut a little smoother when they are slightly dull.

Pitch — In many cases, the tooth pitch (TPI) is dictated by the width and the grind. The width of the blade determines how large the teeth may be, and this, along with the shape of the teeth, determines how closely they can be spaced. Narrow blades have smaller teeth, and therefore have more teeth per inch than wider ones. Blades with skip teeth or hook teeth have fewer teeth per inch than those with standard teeth.

When you have a choice, select the pitch according to how smooth you want the surface and how fast you want to cut. The coarser the pitch (the fewer teeth per inch), the rougher the cut surface; the finer the pitch (the more teeth per inch), the smoother the cut. Blades with a coarser pitch (few teeth) cut faster than those with a fine pitch (many teeth).

FOR BEST RESULTS

If you can vary the speed (fpm) of your band saw, use higher speeds for blades with coarser pitches. This will increase the number of cuts per inch and create a smoother surface. By the same token, if you must cut at a low speed or use an underpowered machine for heavy work, use a blade with a finer pitch — this, too, will increase the cuts per inch.

2-8 Standard band saw teeth are spaced close together and have small gullets; **skip** teeth have roughly the same shape but are spaced much further apart. The rake angle on both standard and skip teeth is normally 0 degrees. **Hook** teeth are spaced like skip teeth, but the rake angle is about 10 degrees and the gullets are much deeper.

You won't always be able to find the perfect blade — the right combination of width, grind, and pitch — for every woodworking operation. In these cases, you must choose a blade whose characteristics match the *most important* requirements of the job.

FOR YOUR INFORMATION

In some cases, the *set style* of the blade (the pattern in which the teeth are set) is also important. Blades with an *alternate* set style cut smoother than other types because more teeth contact the sides of the kerf. A *raker* set produces a rougher surface, but it's a better choice when cutting thick stock because the set style helps to clear the sawdust from the kerf. A *wavy* set is the best choice for cutting metal because the teeth take smaller bites and therefore last longer.

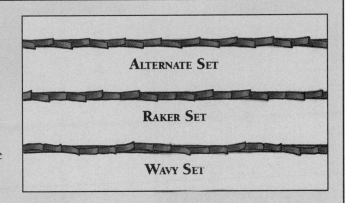

ALTERNATE SET

RAKER SET

WAVY SET

BAND SAW BLADE MATERIALS

The earliest practical band saw blades were made from *spring steel*. This material is still used today for many narrow blades, especially those designed primarily for three-wheel saws. Spring steel is silver in color and fairly soft in comparison with other tool steels. To extend the useful life of the blade, the teeth are tempered with an electric current or flame. However, even spring steel blades with hardened teeth dull quickly.

For this reason, most band saw blades today are made from *carbon steel*. The increased amount of carbon in the steel makes it blacker and harder than spring steel. These blades were originally developed for cutting soft metals, but are now generally used for wood and plastics as well.

Some wide carbon steel blades have *carbide tips* brazed to the teeth. Because the carbide is so much harder than the steel, the teeth stay sharp much longer. In fact, the blade is likely to break long before the teeth grow dull. And carbide is expensive — a ³/₄-inch carbide-tipped band saw blade is almost ten times the cost of a comparable carbon steel blade. A carbide-tipped band saw blade is a good investment only if you do continuous cutting and can braze or weld it when it breaks.

The newest development in blade technology is the introduction of *bimetal* band saw blades. Like carbon steel blades, these were first developed for metal cutting and have found their way to the woodworking market. And like carbide-tipped blades, they are three to four times more expensive than common blades and stay sharp longer. Bimetal blades look like one-piece carbon blades, but the teeth and the body are made from two different metals. A piece of high-speed steel (HSS) is laminated to a strip of carbon steel. The teeth are ground in the brittle HSS edge, leaving most of the flexible carbon steel body.

This type of blade has been touted as an ideal all-purpose blade. However, it has some serious disadvantages. Bimetal blades will withstand a good deal more tension than those made from other materials. Although it's not necessary to overtension these blades, woodworkers who want to resaw with a ¹/₂-inch bimetal blade can tension them the same as a ³/₄-inch carbon blade. Carbon blades and spring steel blades are designed to break when overtightened, but this safety factor is gone with bimetal blades. You could bend the band saw frame, wheel shafts, or other tool parts before the blade breaks. And because it will withstand so much tension, a small bimetal blade may bite into the rubber tires, causing premature wear.

Furthermore, at woodworking speeds bimetal blades fatigue more quickly than other blades. So even though the teeth last longer, the blade body doesn't — in woodworking operations, the bimetal bands break long before the teeth become dull. All this adds up to the fact that bimetal blades just aren't worth the cost if all you're cutting is wood. The most economical blades for the money are the old standbys — spring steel and carbon steel.

(continued) ▷

BAND SAW BLADE MATERIALS — CONTINUED

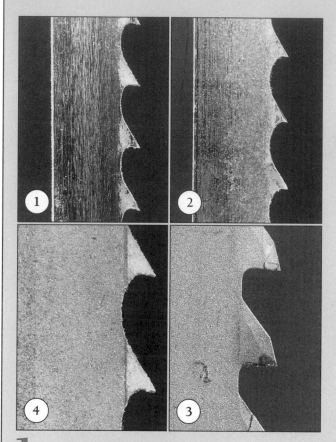

HARD HIGH-SPEED STEEL   FLEXIBLE CARBON STEEL

BEFORE TEETH ARE GROUND

AFTER TEETH ARE GROUND

1 **There are four types of** material commonly used in band saw blades. *Spring steel* (1) is fairly soft but very flexible. The teeth wear out before the blade breaks. *Carbon steel* (2) is harder and stays sharp longer — this is commonly considered the most economical material for blades. Wide carbon blades (¾ inch or wider) are available with *carbide* teeth (3). They hold their cutting edge much longer than any other blade material, but because each blade has so many teeth, the cost of the blade is increased dramatically. Bimetal blades have *high-speed steel* teeth (4) that last longer than either spring steel or carbon steel, though not as long as carbide. They cost more than spring steel or carbon steel blades, though not as much as carbide.

2 **Bimetal blades are made by** laminating a ribbon of hard (but relatively brittle) high-speed steel to a strip of flexible carbon steel. Then teeth are ground in the high-speed steel edge, leaving most of the carbon steel. As a result, the teeth of a bimetal blade are very hard and stay sharp longer, yet the body and back are flexible enough to wear well at low speeds. **Note:** While these blades are not economical for most woodworking operations, they are worth the money if you must cut hard or abrasive materials such as metals, plastics, or particleboard.

COMMON BAND SAW BLADES

BLADE WIDTH	MINIMUM RADIUS*	GRIND	TEETH PER INCH†	USES AND COMMENTS
1/16"	0"	Standard	24	Fine scrollwork. Blade can turn a 90° corner because the kerf is as wide as the blade.
1/8"	3/16"	Standard	14	Cutting scrollwork that requires a smooth surface; also good for joinery
1/8"	3/16"	Skip	8	Fast cutting of scrollwork; cutting scrollwork in thick stock
3/16"	3/8"	Standard	10	Cutting small curves that require a smooth surface; joinery
3/16"	3/8"	Skip	4	Fast cutting of small curves; cutting small curves in thick stock
1/4"	5/8"	Standard	10,14,18	Cutting medium curves that require a smooth surface; joinery. Finer TPIs are best for joinery.
1/4"	5/8"	Skip	4,6	Fast cutting of medium curves in thick stock. For a good general-purpose blade, use 6 TPI.
1/4"	5/8"	Hook	4,6	Aggressive cutting of medium curves; ripping; resawing narrow stock
3/8"	1 1/4"	Standard	8,10,14	Cutting large curves which require a smooth surface; crosscutting and mitering thin stock
3/8"	1 1/4"	Skip	4	Fast cutting of large curves; crosscutting and mitering thick stock. This is also a good general-purpose blade for large saws.
3/8"	1 1/4"	Hook	4,6	Aggressive cutting of large curves; ripping; resawing narrow stock
1/2"	3"	Standard	6,14,18	Cutting gentle curves; crosscutting and mitering. Finer pitches will make very smooth, straight cuts.
1/2"	3"	Skip	4	Fast cutting of gentle curves
1/2"	3"	Hook	4,6	Aggressive cutting of gentle curves; ripping; resawing medium-size stock, sawing green wood. With green wood, 4 TPI works better.
3/4"	5"	Standard	6,8	Crosscutting and mitering thick stock
3/4"	5"	Skip	3,4	Fast cut-off work. May also be used to resaw softwoods.
3/4"	5"	Hook	3,4,6	Aggressive cut-offs; ripping; resawing wide boards; cutting green wood. With green wood, 3 TPI works best.
1"	8"	Standard	8,14	Continuous crosscutting; mitering of thick stock
1"	8"	Hook	3	Continuous ripping, resawing wide boards; cutting green wood

*Sources vary widely on the minimum radius that can be cut with any given band saw blade, and for good reason. The minimum radius depends not only on the width of the blade but also the set and kerf. These can vary from manufacturer to manufacturer. Consider the measurements in this column as estimates only.
†These are the pitches that are commonly available through mail-order suppliers. Other pitches may be available on special order.

BAND SAW SPEED CHANGER

The ideal blade speed varies with the type of material you are cutting, as well as with its thickness and the radius of the cut. The best choice of blade width may also change. For example, when resawing you want to run a wide blade between 2,000 and 3,000 feet per minute (fpm). Small curves are best cut with a narrow blade running around 800 fpm.

Unfortunately, there aren't many variable-speed band saws. However, if you have a belt-driven machine, you can easily make your own speed changer. Instead of connecting the motor pulley directly to the drive pulley, link them via an intermediate arbor. Mount the arbor in ball bearing pillow blocks, bolted to the band saw stand. Secure a four-step pulley on the motor shaft and an *inverted* four-step pulley on one end of the arbor. Connect the two step pulleys with a V-belt. On the opposite end — the power takeoff (pto) — mount an ordinary pulley and connect this to the band saw's drive pulley with a second belt. This setup allows you to run the band saw at four different speeds, depending on how you link the step pulleys.

The motor speed, the diameters of the pulleys, *and* the size of the band saw wheels determine the available speeds. For most woodworking operations, you need a speed range between 650 and 3,100 fpm (approximately). With 4-inch four-step pulleys, you can engineer a speed changer to run a two-wheel band saw in four speed ranges: fast

(2,800 to 3,100 fpm), medium-fast (1,900 to 2,200 fpm), medium-slow (1,100 to 1,250 fpm), and slow (650 to 750 fpm). The *Speed Changer Pulleys* table on the opposite page shows you what pulleys to use depending on the diameter of your band saw wheels. It also shows you what speeds you can obtain with each setup. The *Band Saw Blades and Speeds* table on page 22 suggests the best blade for certain operations and the best speed range to run that blade.

Note: Don't take the information in the *Band Saw Blades and Speeds* table as gospel. Because the table doesn't take into account all the variables that affect each cut, it's meant only as a general guide. If the quality of the cut is paramount, make several test cuts, changing the blade and the speed until you get acceptable results.

1 **To change speeds, move** the position of the belt on the step pulleys. The larger the step on the motor pulley, the faster the band saw will run. **Note:** The belt must link steps that are directly across from one another. Do not run the belt diagonally between the step pulleys.

SPEED CHANGER PULLEYS

BAND SAW WHEEL SIZE	STEP PULLEYS	PTO PULLEY	DRIVE PULLEY	AVAILABLE SPEEDS (IN FPM)*
10″	4″ 4-Step†	3″	8″	3,090; 2,109; 1,218; 747
12″	4″ 4-Step†	2³⁄₄″	9″	2,903; 1,980; 1,144; 702
14″	4″ 4-Step†	2¹⁄₂″	10″	2,814; 1,920; 1,109; 681
14″	4″ 4-Step†	3″	12″	2,853; 1,947; 1,125; 690
16″	4″ 4-Step†	2³⁄₄″	12″	2,965; 2,023; 1,168; 717
18″	4″ 4-Step†	2¹⁄₂″	12″	3,002; 2,048; 1,183; 726

*With a 1,725 RPM motor.
†The sizes of the steps on these pulleys are normally 4″, 3³⁄₈″, 2⁵⁄₈″, and 2″.

2 **To increase the tension of** the belt, let the motor swing down on the motor mount hinge. Then stabilize it by securing the bolts and nuts. Do not use the nuts to add more tension — the weight of the motor should be sufficient. Note that this speed changer uses link-type belts to help eliminate vibration in the power train.

3 **For safety, be sure to make** a guard to cover *all* the pulleys and belts in the speed changer. The guard shown is made from ¹⁄₈-inch and ¹⁄₂-inch plywood, and fastens to the band saw stand with flathead wood screws. The configuration of this guard will change according to the sizes of the pulleys and where you place the motor and inter-mediate arbor.

(continued) ▷

BAND SAW SPEED CHANGER — CONTINUED

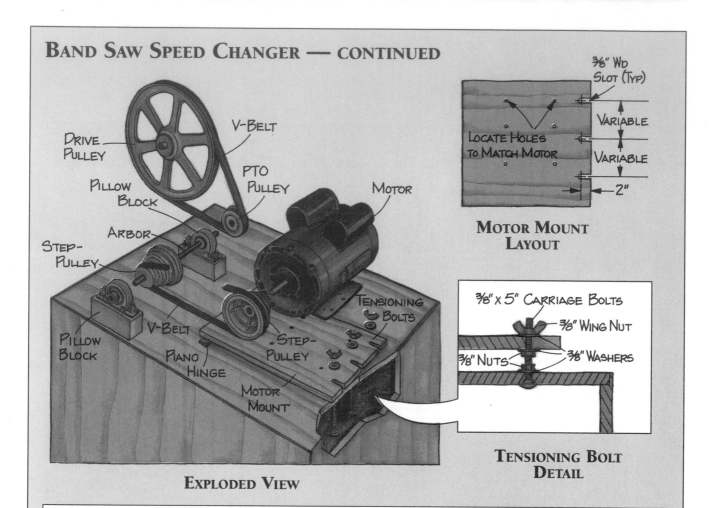

EXPLODED VIEW

MOTOR MOUNT LAYOUT

TENSIONING BOLT DETAIL

BAND SAW BLADES AND SPEEDS

OPERATION (IN WOOD)	RECOMMENDED BLADE WIDTH	RECOMMENDED BLADE TEETH	RECOMMENDED BLADE SPEED
Crosscutting	1/2" to 1"	Standard	Medium-Fast to Fast
Curves, large	1/4" to 3/8"	Skip	Medium-Slow to Medium-Fast
Curves, small	1/16" to 1/8"	Standard	Slow to Medium-Slow
General work	1/4" to 3/8"	Skip	Medium-Fast to Fast
Joinery	1/4" to 3/8"	Standard	Medium-Fast to Fast
Mitering	1/2" to 1"	Standard	Medium-Fast to Fast
Resawing	1/2" to 1"	Hook	Medium-Slow to Medium-Fast
Ripping	1/2" to 1"	Hook	Medium-Slow to Medium-Fast
CUTTING OTHER MATERIALS			
Nonferrous metals	3/8" to 1/2"	Standard	Slow
Paper	3/8" to 1/2"	Standard	Medium-Slow to Medium-Fast
Particleboard	1/4" to 3/8"	Standard	Medium-Slow to Medium-Fast
Plastic	1/4" to 3/8"	Standard or Skip	Slow
Plywood	1/4" to 3/8"	Standard or Skip	Medium-Slow to Medium-Fast

3

CARING FOR THE BAND SAW

Band saw blades require a great deal of support to cut accurately. The blades and bits on most power tools are rigid enough to support themselves and require no additional support other than the arbor or chuck to which they are mounted. Band saw blades, on the other hand, are so extremely flexible that they must be held in tension and braced above and below the work. This prevents them from drifting or twisting as they cut.

To properly support the blades, you must know how to adjust, align, and maintain the apparatus that holds them. The tension and tracking must be properly adjusted for each blade. The blade guides have to be aligned with the blade and with one another. Finally, all the parts that support and guide the blade must be in good working order. None of the alignment or maintenance procedures are difficult, but making them properly requires a basic understanding of how the band saw works, how to keep it working, and what to do when it won't work.

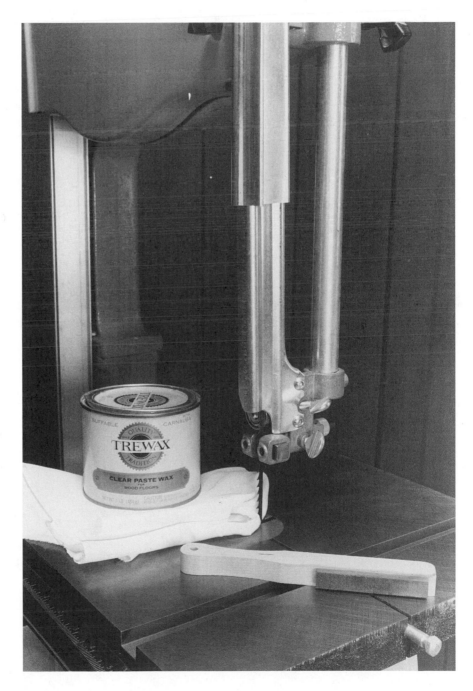

BAND SAW ALIGNMENT AND ADJUSTMENT

Whenever you change a band saw blade, align and adjust the parts of the machine that hold the blade. You may also wish to check the alignment and adjustment if the band saw is not cutting properly or if you want to improve the performance of the blade.

Band saw alignment and adjustment involves five steps:

- Adjust the blade tension.
- Adjust the blade tracking.
- Align the thrust bearings (or plates).
- Align the guide blocks front to back.
- Align the guide blocks side to side.

These procedures do not require a great deal of time, but you must make *all* of them, and in the *proper order.*

ADJUSTING THE BLADE TENSION

Before you begin, retract the thrust bearings and guide blocks so they will not contact the blade. Then loosen the tension knob so you can easily slip the blade over the wheels. Center the blade on the tires and slowly tighten the tension knob until the slack is gone. Rotate the wheels by hand as you continue to tighten the tension knob. (SEE FIGURE 3-1.)

The tension required varies with the width of the blade — the wider the blade, the more tension needed. How do you know when you've arrived at the proper tension? There are several methods, but the most common is to refer to the tension scale on the band saw. Stop turning the knob when the pointer indicates you've arrived at the proper tension for the blade width. (SEE FIGURE 3-2.) Some craftsmen prefer to adjust the tension by sound — they pluck the blade like a giant guitar string, turning the tension knob until the blade produces the loudest, clearest musical sound. Others push against the side of the blade to estimate how it will deflect. When the blade moves about ¼ inch sideways under moderate pressure, the tension is correct.

3-1 Many band saw adjustments must be made with the blade rotating slowly on the wheels. This ensures that the blade is at *equilibrium,* with the tension distributed evenly around the circumference and the body tracking steadily on the tires. However, do *not* make adjustments with the band saw running. Instead, unplug the machine and spin the wheels by hand.

3-2 Tension scales differ from saw to saw, but all of them are calibrated in *blade widths.* The wider the blade, the more tension you must apply. If you apply too much tension to a blade, it will fatigue quickly and break. If you apply too little, it will not cut properly.

If you get the feeling that some tensioning methods are inexact, you're correct — but they still work! Although each blade has an optimum tension at which it will perform best for a given operation, there is a wide range in which it will perform adequately for many operations. Blade tension is important, but you have much latitude as you adjust it.

How do you know when the blade is improperly tensioned? If your blades break long before the teeth dull, you may be applying too much tension. This fatigues the blade body quickly, causing it to snap sooner than it would otherwise. If the blade seems to wander, making it difficult to follow a pattern, or if it cups in the cut when sawing thick stock, the tension may be too loose. *(SEE FIGURE 3-3.)* Usually, you can safely tighten a spring steel or carbon steel blade *one* level past its mark on the tension scale. For example, it's okay to tighten a ¼-inch-wide blade until the scale indicates the proper tension for ⅜-inch-wide blades. The blade may fatigue *slightly* faster than normal, but not overly so. However, if you must raise the tension still further for the blade to cut properly, you have probably chosen the wrong blade for the job.

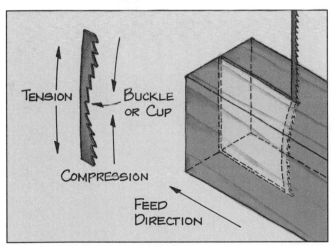

3-3 Tension gives the flexible band saw blade the strength needed to remain straight in the cut. Engineers call this *beam strength*. If the tension is too loose, the blade will bend backward as you feed the wood into it — just like a weak beam bends when you put too much weight on it. This bending compresses the front of the blade and stretches the back, making it buckle or *cup* in the cut. To prevent this, apply more tension to increase the beam strength of the blade.

ADJUSTING THE BLADE TRACKING

Most, but not all, band saws require that you track the blade on the wheels. To do this, *tilt* the tracking wheel as you rotate the wheels by hand. *(SEE FIGURE 3-4.)* The blade must ride over the same place on the tires as the wheels turn — it should not creep forward or back on the wheels. Furthermore, *all* of the blade must remain on the rubber tires. If it rides up on the metal rim of a wheel, the blade or the band saw could be damaged. When the blade is tracking steadily on the rubber, secure the lock. **Note:** A few band saws have automatic tracking devices — a special roller bearing in the column that positions the blade on the tires. Under normal circumstances, you shouldn't have to adjust the tracking on these machines.

Most woodworkers track the blade to ride in the middle of the tires. This gets the blade as far away from the metal rims as possible. However, it's not absolutely necessary. The blade can track toward the front or back edge of a tire without affecting its performance. In fact, band saws with automatic tracking devices track narrow blades toward the back edge of the tires. *(SEE FIGURE 3-5.)*

If your band saw has adjustable tracking, it's best to track the blade so it does not lean forward or back — the blade back should be square to the table. *(SEE FIGURES 3-6 AND 3-7.)* The front-to-back angle between the blade and the table will be *nearly* perpendicular no matter where the blade tracks. But by carefully adjusting the position of the blade on the tires, you can make it precisely perpendicular. When the blade cuts square to the table, the leading edge of the kerf will be straight up and down in the stock. *(SEE FIGURE 3-8.)* This is extremely important when cutting inside corners or making joints.

If you cannot track the blade so it runs square to the table, or if the blade tracks too close to the wheel rims, the band saw wheels may be out of alignment. Realign them *horizontally,* moving one wheel or the other sideways along its axle until the blade tracks in the middle of the tires when it's square to the table. *(SEE FIGURE 3-9.)*

FOR YOUR INFORMATION

Some craftsmen advocate *coplanar tracking* for blades that are ¼ inch wide or wider. To track a blade in this fashion, you must adjust the tracking so the wheels are parallel, then move one wheel or the other along its axle until both wheels rotate in the same plane. The reason given for this procedure is that when the wheels don't turn in the same plane, the stresses on the blade are uneven. This just isn't so — physical laws dictate that the stresses on a running blade always seek equilibrium. If these forces don't strike a balance, the blade won't track and it creeps off the wheels. When a blade tracks — no matter where it tracks or how the wheels are positioned — the stresses are equalized. The prevailing opinion of the tool and blade manufacturers who were consulted for this book is that for any well-made band saw, coplanar tracking isn't necessary. As one engineer put it, "It doesn't do any harm, but it doesn't do any good either. So why do it?"

3-4 To adjust where the blade tracks on the band saw wheels, tilt the idler wheel forward or back. This will move the blade forward or back on the wheels. Although the tilt adjustment is simple, the explanation of what happens when you track a blade is fairly complex: A running blade always seeks the highest spot on a wheel. Tilting changes the position of the crest (if the idler tire is crowned) or shifts the high side (if the tire is flat). Even though you haven't changed the tilt of the drive wheel, the blade will be pulled toward the highest part of the idler wheel. It will seek a point of equilibrium between both wheels, and the track will move forward or back.

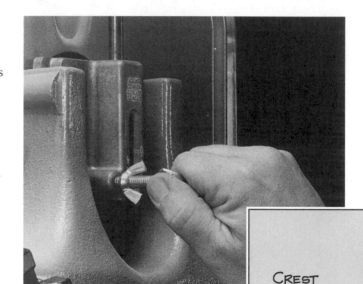

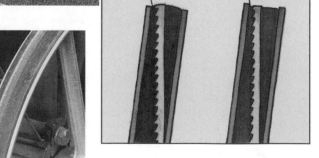

CREST HIGH SIDE

3-5 Band saws with automatic tracking devices are adjusted so the *back* of the blade always tracks in the same place on the tires. Usually, the blade back is well toward the back edge. This means that narrow blades will track off-center. Wider blades will run closer to the center of the wheels.

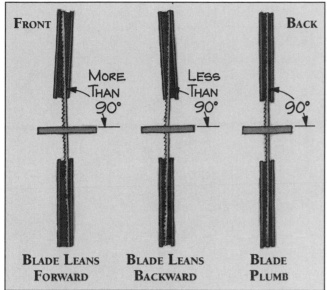

FRONT   BACK

MORE THAN 90°   LESS THAN 90°   90°

BLADE LEANS FORWARD   BLADE LEANS BACKWARD   BLADE PLUMB

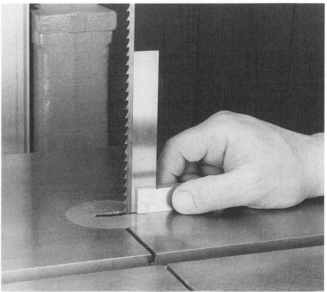

3-6 By adjusting the tracking, you can change the front-to-back angle of the blade a few degrees. When the tracking wheel is tilted so the blade tracks toward the wheel's front edge, the blade leans slightly forward. When it tracks toward the back edge, the blade leans back. Somewhere in between those two positions, the blade will be perfectly plumb — and square to the table.

3-7 To adjust the tracking so the blade is perpendicular to the table, rest a small square on the band saw table so it contacts the back of the blade. While slowly rotating the band saw wheels, tilt the tracking wheel forward or back until the blade back is parallel to the square. As you make this adjustment, be careful *not* to let the blade track too close to the edges of the wheels. If the blade comes any closer than $1/8$ inch to the wheel rims, you may have to adjust the horizontal position of the wheels.

3-8 If you make multiple cuts that meet in the interior of a board, the blade must be square to the table. If the blade leans forward, the cuts won't meet on the bottom of the board even though they meet on the top. If the blade leans backward, the cuts will intersect and go past one another on the bottom of the board, although they just barely meet on the top.

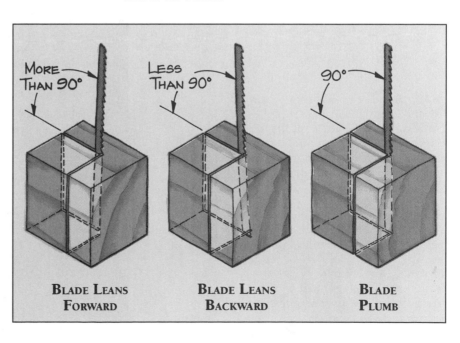

MORE THAN 90°   LESS THAN 90°   90°

BLADE LEANS FORWARD   BLADE LEANS BACKWARD   BLADE PLUMB

3-9 **To align the wheels** horizontally, mount the blade that you use *most often* on the band saw. (For most craftsmen this will be a ¼ or ⅜-inch-wide blade.) Track the blade in the middle of the wheels and check whether the blade leans forward or back. Remove the blade from the saw and move one wheel or the other along its axle so the blade will run plumb. On some band saws, you must loosen a set screw to move a wheel; on others, you must remove the axle nut and insert shims or washers between the frame and the wheel (as shown). Replace the blade, and check the front-to-back angle. You may have to go through this procedure several times to get the blade square to the table.

ALIGNING THE THRUST BEARINGS

With the blade tensioned and tracking properly, rotate the blade until the weld is just in front of the upper blade guide. Advance the upper thrust bearing (or plate) until it just touches the weld. Then back it off the thickness of a piece of paper. *(See Figure 3-10.)* Reposition the weld in front of the lower blade guide and repeat.

   Why use the blade weld? Because if there is a kink in the blade so the blade comes a fraction of an inch closer to the thrust bearings, the weld is likely to be it. The weld is also a handy reference spot on the blade, allowing you to adjust both the upper and lower thrust bearings precisely the same. It's extremely important that these bearings be aligned vertically. If one of them leads the other by even a small amount, they won't support the blade equally. This, in turn, may cause the blade to twist and make it difficult to saw a straight, even cut.

3-10 **To position a thrust bearing,** advance it until it just touches the blade weld, then back it off the thickness of a piece of paper. Do the same for the lower thrust bearing. Rotate the blade by hand, watching the thrust bearings closely. If the blade touches them and causes them to turn, back them off another paper's thickness. Be careful to move *both* bearings the same amount. A thrust bearing must not touch the running blade *unless it's cutting*. In addition, both thrust bearings must be aligned vertically to hold the running blade straight up and down.

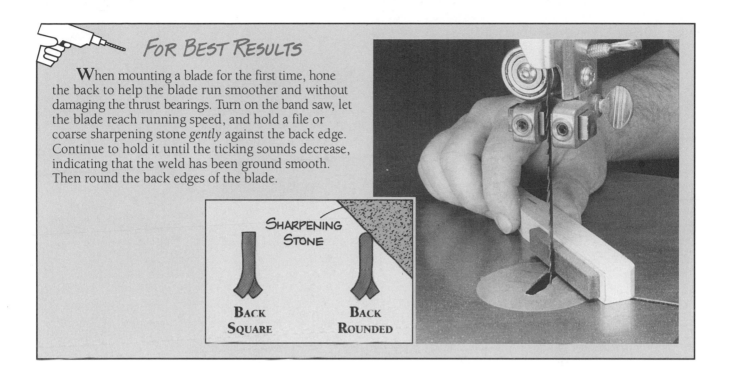

FOR BEST RESULTS

When mounting a blade for the first time, hone the back to help the blade run smoother and without damaging the thrust bearings. Turn on the band saw, let the blade reach running speed, and hold a file or coarse sharpening stone *gently* against the back edge. Continue to hold it until the ticking sounds decrease, indicating that the weld has been ground smooth. Then round the back edges of the blade.

SHARPENING STONE

BACK SQUARE

BACK ROUNDED

ALIGNING THE GUIDE BLOCKS FRONT TO BACK

The guide blocks serve two important functions. First, they prevent the blade from twisting as you cut. Second, they keep the blade running true, damping out vibrations caused by slightly out-of-round wheels and other small imperfections in the machine or the setup.

To position the guide blocks, advance the upper and lower blade guides until the front edge of each block is even with the bottoms of the gullets between the blade teeth. (*SEE FIGURE 3-11.*) This prevents the guide blocks from damaging the teeth — and the teeth from damaging the guide blocks!

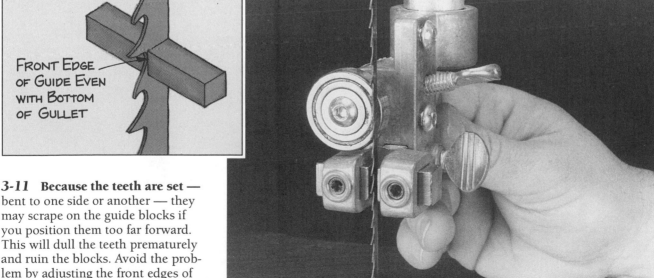

FRONT EDGE OF GUIDE EVEN WITH BOTTOM OF GULLET

3-11 **Because the teeth are set —** bent to one side or another — they may scrape on the guide blocks if you position them too far forward. This will dull the teeth prematurely and ruin the blocks. Avoid the problem by adjusting the front edges of the guide blocks even with the bottoms of the blade gullets.

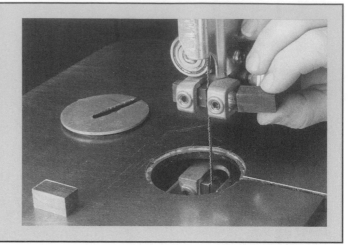

ALIGNING THE GUIDE BLOCKS SIDE TO SIDE

Advance the upper and lower guide blocks on one side of the blade (either right or left) so they just touch the blade body, then back them off the width of a piece of paper. (SEE FIGURE 3-12.) Do the same for the blocks on the other side.

When you have completed all five steps and the band saw is properly adjusted to run the blade, double-check that all the set screws, nuts, and knobs are secured. Also replace the leveling pin (or table tie) and the blade insert.

3-12 Position the upper and lower guide blocks on one side of the band saw a paper's thickness away from the blade. Tighten the set screws that hold the blocks in place, and position the blocks on the other side the same way. Rotate the blade by hand to make sure it doesn't contact the guide blocks at any point. If it does, back off both the upper and lower guide blocks (on the side that contacts the blade) the same amount. Like the thrust bearings, the running blade must not touch the guide blocks *until it cuts.* Also, the upper and lower blade guides must be aligned vertically. If they aren't, the blade may be deflected to one side or the other.

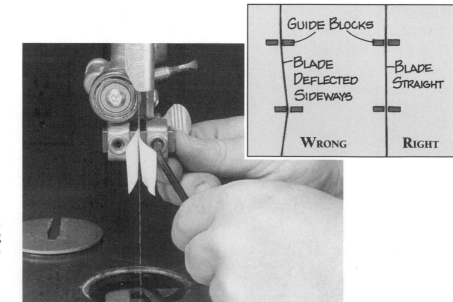

BAND SAW MAINTENANCE

MAINTAINING THE TOOL

For a complex tool, a band saw requires relatively simple maintenance. You must, of course, keep the machine properly aligned and adjusted. Not only does this help the band saw to perform better, it cuts down on wear and tear of the components. In addition, there are chores you should tend to periodically.

Cleaning — Next to proper alignment and adjustment, this is the most important task you can perform to keep the band saw running smoothly. Working on the band saw generates *lots* of fine sawdust — more dust than some sanding operations! Although it seems relatively benign, sawdust is a fine abrasive that can ruin the working parts of a power tool if allowed to accumulate. It gets into wheel bearings and other working parts, causing premature wear. It also becomes imbedded in the tires, interfering with the running blade.

To prevent sawdust buildup, outfit your band saw with a dust collector and use it every time you turn on the machine. (*SEE FIGURE 3-13.*) Now and then,

remove the cover and vacuum the sawdust from all the crevices in the frame. Also, clean the wheels and tires. (*SEE FIGURE 3-14.*) Whatever your owner's manual recommends, do it every time you clean the shop. While you have the shop vacuum out, it takes only a few extra minutes to clean the tools.

FOR BEST RESULTS

Whenever you cut plastic or metal on a band saw, you should clean the machine before using it to cut wood. These materials may become embedded in the tires, causing them to wear prematurely. Additionally, the machine may throw metal filings and plastic shavings onto the wood. These scratch the surface and may interfere with some finishes.

Waxing and lubrication — Band saws with permanently sealed wheel bearings need no lubrication. If you keep these bearings clean, they should last for years. The wheels of some inexpensive saws are mounted on bushings, which do require periodic lubrication. However, don't use oil or grease unless

3-13 Your band saw will run better with less wear if you vacuum the sawdust as you work. If your band saw doesn't have a built-in dust collector, cut a 2-inch-diameter hole in the bottom cover near the rim of the wheel. Fasten a collar over this hole to attach a vacuum hose. This will remove about 70 percent of the sawdust as it's generated. For a more efficient dust collector, attach a second hose as close as possible to the lower blade guide.

3-14 To remove the sawdust and pitch that builds up on the band saw tires, scour them with #3 steel wool or a plastic scouring pad. With the band saw turned *off*, turn the wheels by hand while you hold the pad against the tires. Do *not* attempt to do this with the band saw running.

specified in the owner's manual. These materials mix with fine sawdust, forming an abrasive goo that may increase the wear on the working parts. Instead, use a dry lubricant such as silicone or graphite.

Every few months (or every few weeks, if you use your band saw constantly), apply paste wax to the worktable and buff it. This will help the work to slide smoothly across the surface. It also helps protect the table from corrosion. **Note:** You must buff out the wax thoroughly! Once buffed, the thin layer of polished wax lubricates the surface. But if you leave too much wax on the table, it will dry to a gummy paste that has exactly the opposite effect. Furthermore, the excess wax will contaminate the surface of the wood that passes over it. This, in turn, will interfere with some finishes.

Resurfacing the guide blocks — Whether you have guide blocks of metal or some other material, sooner or later they will become worn. The deeper the wear marks, the less support the guides will provide for the blades. When you notice a guide block is scored or uneven, grind the surface smooth on a stationary sander or sharpening stone. (*SEE FIGURES 3-15 AND 3-16.*)

FOR YOUR INFORMATION

Although you can change your own band saw tires, it's not advisable. The rubber must be stretched very evenly around the rim or the wheels will be out of round. You'll get better results if you send your wheels to a repair shop or back to the manufacturer, where they have the equipment to do this.

Resurfacing and replacing the tires — Over a long period of time, the surface of the tires hardens. To restore the resilience of the rubber surface, dress the tires by sanding them with 100- or 120-grit sandpaper. (*SEE FIGURES 3-17 AND 3-18.*) Eventually, the rubber will harden clear through, and the tires will crack and split. When this happens, you must have the tires replaced.

3-15 You can resurface a guide block on a strip sander (shown), belt sander, or stationary sander. However, you will have to use a miter gauge or devise a guide to grind the end of the block at the proper angle. (The ends of most guide blocks must be ground square to the sides. A few are ground at 45 degrees.) Use 100- or 120-grit sandpaper to remove the score marks left by the saw blades, then polish the surface with 220-grit or finer.

3-16 You can also resurface the end of a guide block with a sharpening stone, but you must devise a guide to hold the block at the proper angle. Here, a block is cradled in an L-shaped jig made from two scraps of wood. This holds the block square to the stone.

3-17 To dress the band saw's drive wheel tire, open the cover and remove the blade. Cover a cloth pad or a wad of steel wool with 100 or 120-grit sandpaper — this will conform to the shape of the wheel. Turn on the motor and hold the sanding pad *lightly* against the rubber for a *few moments* — just long enough to remove the wood pitch and the glaze of the hardened rubber. Test the surface of the rubber with your thumbnail to make sure that the tire has some give. If it's hardened all the way through, it must be replaced.

3-18 To dress the idler tire, rotate the wheel with a power drill and a large drum sander. (Do *not* mount a blade and use it to turn the idler wheel — this can be *very* dangerous.) Have a helper hold the sanding drum against the tire and turn on the drill. As the idler wheel gets up to speed, lightly sand its tire in the same manner that you sanded the drive wheel tire.

CLEANING AND REPLACING BLADES

The most important task in maintaining a blade is to keep it clean. As the blade cuts, wood pitch builds up on the cutting face and in the gullets. If allowed to accumulate for very long, this pitch interferes with the cutting action of the teeth. It also increases the friction on the running blade, causing it to heat up and break prematurely. To prevent these problems, clean the blade with oven cleaner. Naphtha, mineral spirits, and some other solvents will also dissolve the pitch. (*SEE FIGURE 3-19.*)

When blades become dull they should be replaced. Not too long ago, it made good economic sense to have band saw blades sharpened once or twice before discarding them. In recent years, however, the cost of

sharpening a blade has risen while the price of blade stock has become more reasonable. It now costs less for a saw sharpening service to *make* a new blade than it does to sharpen an old one. Many shops have stopped sharpening band saw blades altogether.

A BIT OF ADVICE

Often, a blade that seems dull is simply covered with pitch. Try cleaning the blade before you discard it.

If you use a lot of band saw blades, you can make your own from blade stock for about half the price of buying them. This requires a special clamp to hold the ends of the blade stock together while you braze them with silver. (*SEE FIGURE 3-20.*) A silver braze, when properly done, is a strong, flexible joint that will endure almost as long as a weld.

> ## TRY THIS TRICK
>
> **O**ne of the advantages of being able to splice your own blades is that you can make *interior* cuts. Snip the blade with bolt cutters and thread it through a hole in the workpiece. Then braze it back together and mount it on the band saw. Cut out the interior shape, relieve the blade tension, and snip the blade again to remove it from the stock. This technique is more often used in large furniture factories than small shops, where it's customary to make interior cuts in wood with a scroll saw or saber saw. But depending on the cut you have to make, it may be worth the effort.

3-19 To clean a blade, you must dissolve the pitch that builds up between the teeth. Of the many solvents that will do this, one of the quickest is oven cleaner. Remove the blade from the saw, fold it, and place it in an old pan or on a sheet of plastic. Spray the blade thoroughly with oven cleaner — the blade does *not* have to be warm for the cleaner to work. Wait a few moments for the pitch to dissolve, then unfold the blade and wipe it with a rag. Draw the blade through the rag *backward*, so the teeth won't snag.

3-20 To make your own blade, cut the proper length from a roll of blade stock. Bevel the ends with a file at 20 or 25 degrees and clamp them together so the back of the blade will be straight and flush. Heat the blade with a propane torch and apply silver solder to the joint. Blade stock, clamps, and solder are sold through many mail-order woodworking supply companies.

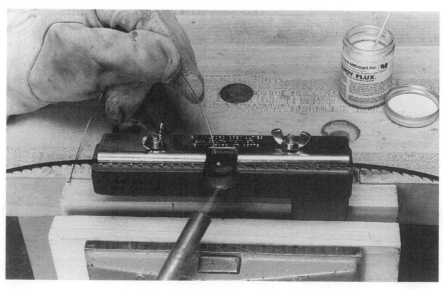

FOLDING BAND SAW BLADES

Commercial band saw blades usually come folded in a triple loop. Many manufacturers caution against refolding a blade once you have used it, because this fatigues the blade and may shorten its useful life.

However, few craftsmen have the storage space to keep unfolded band saw blades hanging around. Although it fatigues the blade slightly, it's more convenient to fold a blade to store it. How do you fold a band saw blade? That's a simple trick — twist the blade one full turn and it will fold itself.

2 **Turn your hands in** opposite directions, twisting the blade so the "thumbs-up" hand is "thumbs-down" and vice versa.

1 **Grasp the blade with one** hand at the 3 o'clock position on the loop and the other hand at 9 o'clock. Turn your hands so one palm faces out and the other faces in. One hand will be in a "thumbs-up" position; the other will be "thumbs-down."

3 **Bring your hands together.** At this point, the blade will naturally fall into three equal coils.

4

BAND SAW KNOW-HOW

There are two ways to make a cut on a band saw. You can saw *freehand*, controlling the cut by turning the board this way and that as you feed it past the blade. Or, you can make *guided cuts*, using a miter gauge, fence, or some other jig to guide the workpiece as it's sawed.

Most band saw cuts are made freehand, and for a good reason — it's the simplest way to use the machine. When cutting freehand, you can easily compensate for blade lead and other idiosyncrasies of the tool by changing the angle at which you feed the workpiece. As long as you choose the proper blade for the job and take your time when cutting, you can get good results with surprisingly little effort.

Guided cuts require more time and finesse. Unlike freehand cutting, you cannot adjust the feed direction during a guided cut, so you must anticipate how the blade will cut and align the fence or miter gauge accordingly. Proper alignment requires you to make experimental cuts in scraps and measure the results. When making the cut, you must feed the stock carefully and steadily. This makes a guided cut more work than a freehand cut; however, the results are more consistent and accurate.

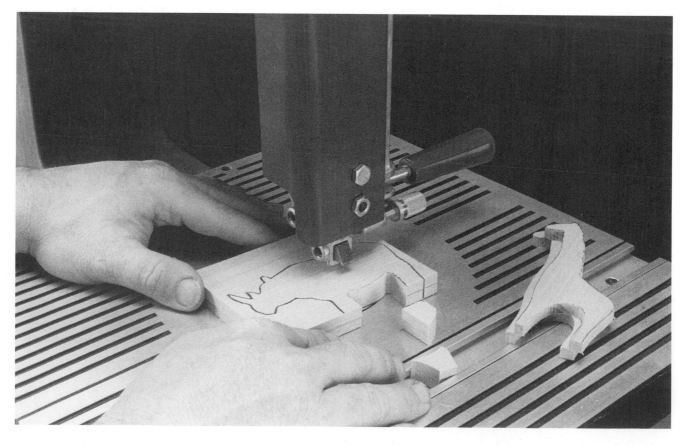

ESSENTIAL CUTTING TECHNIQUES

Whether you cut freehand or with a guide, there are some common procedures you must follow for good results.

BEFORE YOU TURN ON THE SAW

As you get ready to cut, verify that the table is at the proper side-to-side angle to the blade with a small square or a protractor. (*SEE FIGURE 4-1.*) Also, adjust the height of the upper blade guide so it clears the work by just 1/8 to 1/4 inch. (*SEE FIGURE 4-2.*)

A SAFETY REMINDER

Never use the band saw with the upper blade guide raised more than 1/4 inch above the work. The portion of the blade between the guide and the work is not covered by the blade guard. The more that shows, the greater the chances for an accident.

Think through the cut. Lay out the pattern so the bulk of the workpiece will be *outside* the blade, not caught in the band saw's throat between the blade and the column. (*SEE FIGURE 4-3.*) Mark the waste side of the layout lines so you know precisely what stock to remove. This is especially important when cutting complex patterns. It requires lots of concentration to monitor the blade and adjust the feed angle — so much so that after making a few turns, you can easily lose track of what's scrap and what isn't.

Also make sure the work is properly supported. Band saw tables are small — smaller than many workpieces. If you find it difficult to keep the stock flat on the table as you cut, use a saw stand for additional support. (*SEE FIGURE 4-4.*) Or, attach an auxiliary table to your band saw to extend the work surface. "Auxiliary Band Saw Tables" on page 47 gives instructions and plans for two different sizes of tables.

4-1   Every time you change a blade or the table tilt, check that the blade is at the proper side-to-side angle to the table *before* you make a cut. A small engineer's square works well when confirming that the blade and table are square. To verify other angles, use a small protractor with a square head (shown). Raise the upper blade guide and rest the arm of the square or protractor against the side of the blade body. Make sure it doesn't touch the teeth.

4-2   The upper blade guide must clear the work by no more than 1/8 to 1/4 inch. There are two reasons for this. First, it's not safe to leave more than 1/4 inch of blade exposed — your finger could accidentally slip under the guide and into the blade. Second, the closer the upper guide is to the lower one, the better the blade will be supported in the cut. When properly supported, the blade is less likely to vibrate or twist, and it's easier to control the cut.

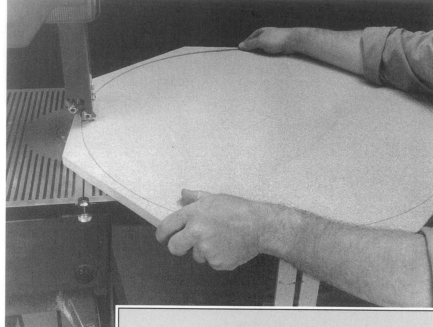

4-3 When sawing pieces that are larger than the throat of your band saw, plan your cuts so that most of the stock will remain outside the throat. If necessary, trim the waste with a handsaw or saber saw so it won't hit the column as you cut.

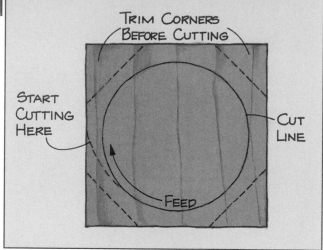

4-4 If the work is too long or too large for the worktable, use one or more saw stands to provide additional support. Place the stands where they will support the bulk of the workpiece while being cut.

AS YOU CUT

The band saw is one of the safer power tools in your shop because the cutting action of the blade helps hold the work on the table. This eliminates the danger of kickback, but it doesn't mean you can let down your guard. Any machine that cuts wood will cut you.

When cutting on the band saw, the "danger zone" extends 1 to 2 inches out from the blade. You should never allow any part of your body to stray into this area. If something unforeseen should happen while your hands are inside the danger zone, you may not be able to react fast enough to retrieve them without injury. For this reason, use push sticks and other safety devices to manipulate wood close to the blade.

TRY THIS TRICK

To safely cut a small piece of wood on the band saw, adhere it to a larger scrap with double-faced carpet tape. Cut both pieces, manipulating the smaller board with the larger one. This will keep your fingers out of the danger zone.

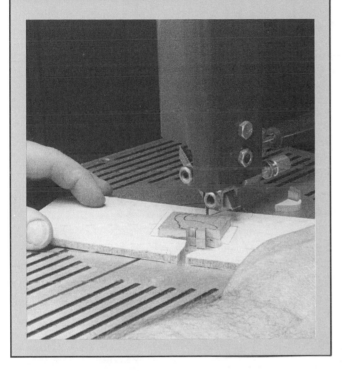

Feed the wood into the blade, cutting on the waste side of the layout line. Turn the workpiece as necessary, adjusting the feed direction so the blade follows the line. A slow, steady speed usually produces the smoothest possible cut. Be careful not to feed the wood so fast that the machine bogs down.

If the line is hard to follow or the blade wanders or drifts in the cut, several things may be wrong:

■ The blade width may be too large to follow the contours of the pattern. Change to a smaller blade or use one of the techniques for cutting small curves with a large blade described in "Making Freehand Cuts" on page 41.

■ The blade may be too loose. Apply more tension and, after you do so, check the tracking and blade guides.

■ The blade guides may not be properly adjusted. Check the thrust bearings and guide blocks and, if necessary, reposition them. Also check that the upper blade guide is adjusted just above the work.

■ The blade may be dull or clogged with pitch. First, try cleaning the blade. (Refer to "Band Saw Maintenance" on page 31 for instructions.) If that does no good, change to a sharper blade.

Break complex patterns into short, simple cuts. Think each operation through and plan the sequence of cuts that will produce the contours you want. Oftentimes, you must remove the waste in several steps. (SEE FIGURE 4-5.) If an operation requires two or more band saw cuts to remove a piece of the waste,

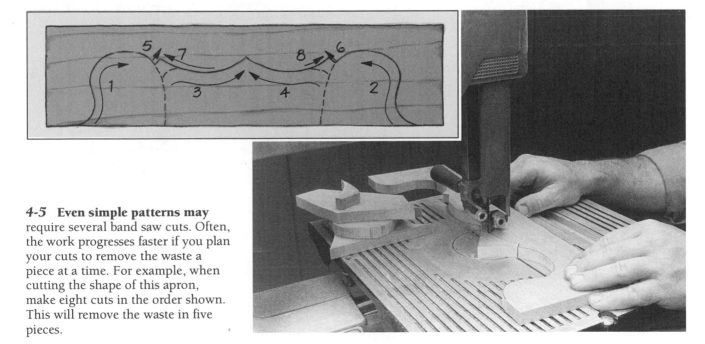

4-5 **Even simple patterns may** require several band saw cuts. Often, the work progresses faster if you plan your cuts to remove the waste a piece at a time. For example, when cutting the shape of this apron, make eight cuts in the order shown. This will remove the waste in five pieces.

make the short cuts first, if possible. (*See Figure 4-6.*) This will reduce the amount of time you spend backing the blade out of kerfs.

If it's important that the cut surface be as smooth as possible, turn off the band saw when backing out of cuts. Otherwise, the teeth may dig into the sides of the kerf. It's also advisable to turn off the saw when backtracking sharp turns. This reduces the risk that you may twist the blade or pull it off the wheels.

Try This Trick

To reduce the amount of backtracking required to cut a pattern, drill holes in the waste at strategic locations. The diameter of these holes must be a little larger than the width of the blade so you can change the direction of the cut to any angle. For example, when making the long V-shape shown, drill a hole in the waste near the point of the V. Cut one side of the V, backtrack a short distance, and cut to the hole.

Turn the workpiece and saw toward the edge, cutting as much of the second side of the V as possible. Finally, go back and cut the last portion of the second side, cutting toward the point of the V.

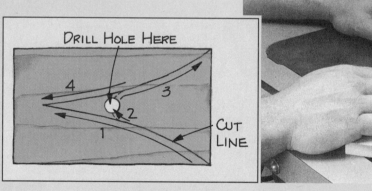

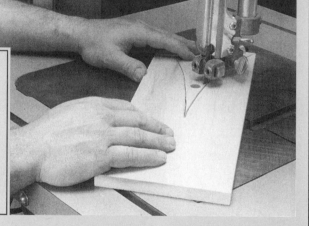

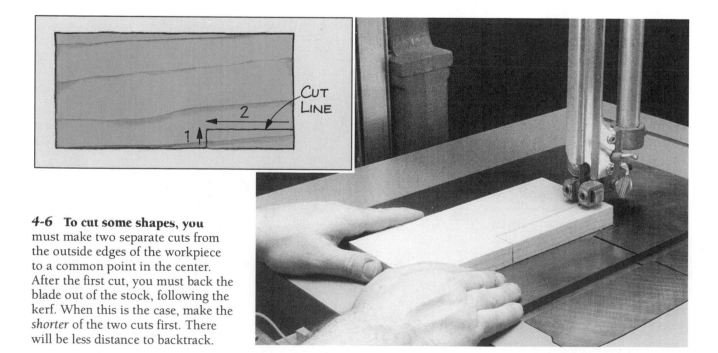

4-6 To cut some shapes, you must make two separate cuts from the outside edges of the workpiece to a common point in the center. After the first cut, you must back the blade out of the stock, following the kerf. When this is the case, make the *shorter* of the two cuts first. There will be less distance to backtrack.

FINISHING THE CUT

This is the most dangerous part of any band saw operation. For most of the cut, the blade is safely embedded in the wood and you don't have to worry much about the danger. But when the blade is ready to exit, you must pay close attention. If you misplace your hands, the emerging blade could easily cut off a finger.

For this reason, it's *always* a good idea to finish the cut by feeding the stock past the blade with a push stick, push block, or scrap of wood. (*SEE FIGURE 4-7.*) If the stock is small or narrow and you cannot finish the cut without putting your hands in the danger zone, you *must* use a safety tool.

TRY THIS TRICK

Paint the band saw table insert bright red, both as a reminder of the danger zone and to warn yourself when the blade is about to exit the wood. During most of the cut, the wood covers the insert. But toward the end, the insert appears. When you see red, you know it's time to reach for a push stick.

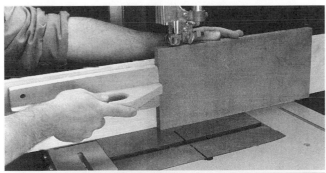

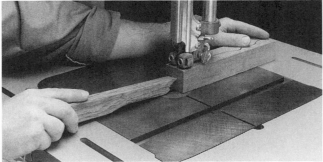

4-7 Although you may not need a safety tool at the beginning of a band saw cut, it's always a good idea to use one when finishing up. This prevents you from accidentally putting your hand in the path of the blade.

After the blade exits the wood, turn off the motor and let the machine stop completely *before* removing small pieces from around the blade. This is especially important with two-wheel band saws. The large wheels have a lot of momentum — "stored kinetic energy," physicists call it. Even though the motor is off, there is enough of this energy behind the coasting blade to cut through an inch or two of wood — or flesh — before it comes to rest.

MAKING FREEHAND CUTS

It's simple to get good results when cutting freehand on the band saw — just feed the work slowly enough to give yourself time to react when the blade begins to drift away from the layout line. And remember: Don't just follow the line — *shave* it. Cut to the waste side so the blade just grazes the line. When you're finished cutting, most of the pencil or the score mark should remain on the wood. This isn't as difficult as it sounds, although it requires a little practice to get the knack.

FOR BEST RESULTS

To make accurate freehand cuts, you must have *good lighting*. You can't shave a line if you can't see it. Some band saws come with built-in illumination, but if yours doesn't, place a light stand or tension lamp nearby, as shown. Use *incandescent* bulbs to light the worktable. Fluorescent bulbs flicker with the oscillations in the electric current, faster than your eye can detect. If the band saw blade travels at a synchronized speed with these oscillations, the blade may look as if it's standing still — even though it's moving! *This can be very dangerous.*

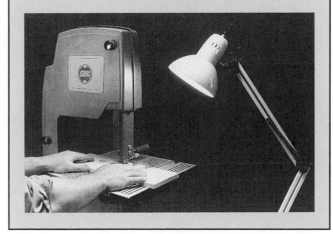

CUTTING CURVES

Most freehand cuts are curved cuts. These present no special problems as long as the curve radii are larger than the minimum radius that can be cut by the blade. (Many band saw owners find that gentle curves are actually easier to cut than straight lines.) But what if the radius of a curve is too small for the blade? You have two choices: You can either change to a narrower blade, or you can employ one of these special techniques:

Making tangential cuts — Oftentimes, you can cut a small outside curve with a large blade by breaking it up into a series of cuts. Cut as much of the curve as possible without twisting the blade — you'll find you can follow the layout line for a short distance until the blade begins to drift from the pattern. As soon as this happens, cut a straight line tangent to the curve through the waste to the outside edge of the stock. (*SEE FIGURE*

4-8.) This will free a portion of waste. Remove it, then go back and cut another portion of the curve, beginning where the blade first started to drift from the layout line. If the blade drifts again, make another tangential cut. Continue until you have completed the curve.

Making radial cuts — If you know an outside cut is too small for a blade, make several cuts radial to the curve from the outside edge through the waste to the pattern line. This will divide the waste into several segments. Then cut the curve, removing the pieces of waste as you do so. (*SEE FIGURE 4-9.*) The blade won't wander from the curve because the back of the blade never rubs against the waste side of the kerf. And it never rubs because the waste is removed a piece at a time as you cut.

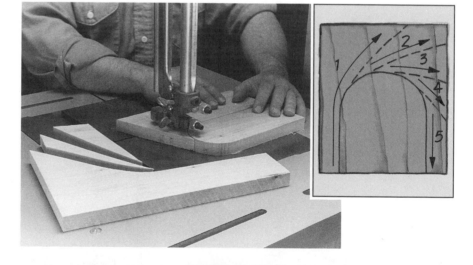

4-8 If the blade begins to drift from the cut when cutting a tight curve, don't try to force it to follow the layout line. You will twist the blade, causing its back to rub in the cut. This, in turn, will heat up the blade and burn the wood. Instead, allow the blade to drift into the waste, cutting a line *tangent to the curve*. Remove a portion of the waste, return to where the blade left the pattern, and cut again. Make as many of these tangential cuts as needed to complete the cut.

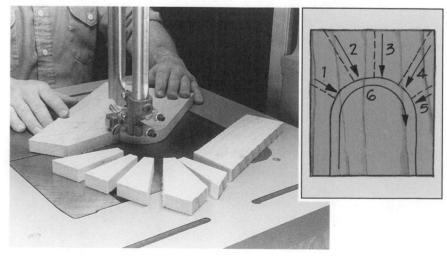

4-9 Another way to cut a tight outside curve is to make several *radial cuts*, cutting from the outside edge of the workpiece to the pattern line prior to cutting the curve. Ideally, these cuts should be spaced so they're no further apart than the width of the blade where they meet the pattern. This will divide the waste into small segments that will be removed as you cut. Just as the blade begins to rub on the side of the kerf, you'll reach one of the radial cuts. A portion of the waste will separate, and the blade will no longer rub.

Enlarging the kerf — You can also cut tight outside curves by enlarging the kerf when the blade begins to rub. Cut until the blade starts to drift from the pattern line, then backtrack a short distance in the cut. Cut out into the waste slightly, widening the kerf where the blade left the pattern. (*See Figure 4-10.*) This will stop the back of the blade from rubbing in the kerf and enable you to keep cutting. Cut until the blade drifts again, then repeat.

Drilling relief holes — To cut a tight *inside* curve, you can drill a hole in the waste the same radius as the curve. This is sometimes referred to as a relief hole. You must position the drill carefully so the curved edge of the relief hole follows the pattern line without removing any of the good stock. (*See Figure 4-11.*) If necessary, use a compass to locate the center of the hole.

4-10 You can also *enlarge the kerf* to cut a tight outside curve. When the blade begins to drift, backtrack in the cut two or three times the width of the blade. (For example, if you are working with a ¼-inch-wide blade, backtrack ½ to ¾ inch.) Shave the waste side of the kerf, making it wider. Depending on how tight the cut is, you may have to shave it several times to enlarge the kerf to the dimensions needed. Then continue cutting the curve. Like the radial cut technique, this method keeps the back of the blade from rubbing on the side of the kerf.

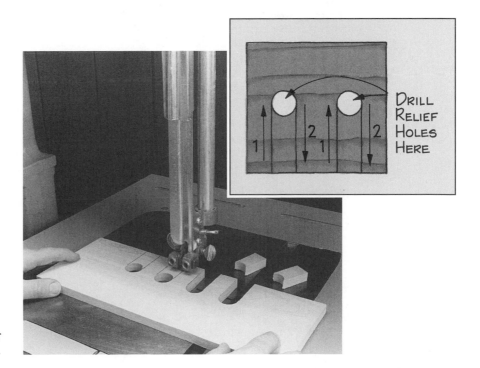

4-11 You can cut tight inside curves by *drilling relief holes* in the waste. The radii of the holes and the curves in the pattern should match. If you want the curved surface to be as smooth as possible, use a Forstner bit or a multispur bit to drill the holes.

Nibbling — Finally, you can cut away the waste from any tight curve, inside or outside, by nibbling at it with the blade. Instead of cutting a narrow kerf, use the teeth on the blade like a file to remove a small amount of stock from a broad surface. *(SEE FIGURE 4-12.)* This technique requires a very gentle touch — you must feed the wood very slowly and not try to remove too much stock at once.

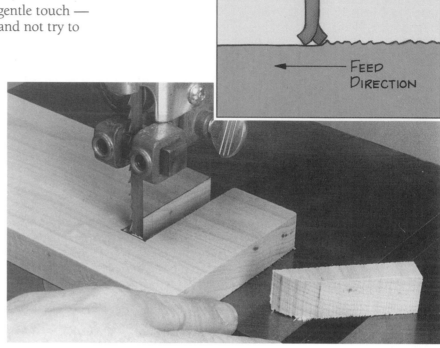

4-12 To use a band saw blade to *nibble* away at the stock, you must cut sideways with the blade. Instead of feeding the wood into the blade parallel to the blade body, feed it *perpendicular* to the blade body. Remove just a shallow layer of stock, about $1/32$ inch deep. If you try to remove too much, the blade won't cut. Continue nibbling until you cut up to the pattern line. With a little practice, you can make a nibbled surface as smooth as one that has been cut conventionally. This technique is useful for cutting small curves and the bottoms of deep slots. *(Blade guide raised for clarity.)*

MAKING GUIDED CUTS

FINDING THE LEAD LINE

When using a fence or miter gauge to guide a band saw cut, the results depend on the accuracy of your setup. You cannot change the feed direction as you can when cutting freehand, so adjust the blade to cut steadily in a straight line, with no tendency to wander or drift. To do this, set the tension, tracking, and blade guides properly. In addition, *the feed direction must be parallel to the blade lead.*

Some manuals tell you to adjust a fence or miter gauge slot parallel to the body of the blade. This is nearly impossible, since band saw blades are not wide enough to make the adjustment accurately without special measuring tools. It's also useless. Unlike rigid blades (such as circular saws), a band saw blade doesn't always cut parallel to its body. Furthermore, one band saw blade won't cut in the same plane as another. When you change blades, the new blade may lead a few degrees to one side or the other, relative to the old one. Finally, the blade lead may change slightly as the blade dulls.

There are many factors that affect blade lead. The teeth of many inexpensive blades are punched from steel bands, creating a burr on one side — and the blades lead toward the side without the burr. If the teeth have nicked the blade guides, dulling the blade on one side, the blade will lead toward the other side — sharp teeth cut more aggressively than dull ones. If the teeth are set unevenly or something has happened to knock the set off the teeth on one side, the blade will lead toward the side with the greatest set. In general, the direction of blade lead depends on the condition of the teeth — their relative sharpness and set.

How do you compensate for blade lead? Some manuals advise you to hold a sharpening stone lightly against the "lead side" of the running blade. This removes some of the set and dulls the teeth on one side so it no longer cuts more aggressively than the other. This brings the blade lead more in line with the blade body. Unfortunately, it also shortens the useful life of the blade.

It's much easier on your blade to readjust a fence or miter gauge parallel to the blade lead each time it becomes necessary. To check the blade lead, first make sure the blade is tensioned, tracked, and guided properly. Then select a long, straight wood scrap and make a test cut to see which direction you must feed the wood to cut a straight line. (*See Figure 4-13.*) Using the scrap like a straightedge, mark a line that indicates this direction on the worktable. (*See Figure 4-14.*) When making guided cuts, feed the wood parallel to this *lead line*.

TRY THIS TRICK

When drawing lead lines on your worktable, mark both sides of your test piece so you have lines on *both* sides of the blade. This will make it easier to adjust fences and other guides. If you need to position the guide on the worktable so it covers one lead line, the other one will still show.

USING FENCES AND MITER GAUGES

When using a fence, you must adjust it parallel to the lead line. (*See Figure 4-15.*) A good band saw fence should be designed to do this quickly. If the fences that are available to fit your saw can't be adjusted in such a way, make your own. Refer to the plans and instructions for "Resawing Aids" on page 61.

Adjusting the miter gauge slot is much more difficult. It's a simple matter to loosen the bolts that hold the table to the trunnions and shift it clockwise or counterclockwise. However, you can't use the lead line to tell you how far to turn the table because the line turns with the table! So you must find the proper table position by trial and error.

For this reason, the miter gauge isn't used much on the band saw. Most craftsmen rely on it only for operations where the alignment isn't critical, such as cutting off narrow boards or small dowels. (*See Figure 4-16.*) **Note:** You can cut with the miter gauge when the slot is out of alignment by a few degrees. However, you must feed the stock slowly, giving the blade time to cut a straight line even though the feed direction isn't quite correct. The greater the misalignment, the slower you should feed the wood.

4-13 To find the blade lead, make a test cut in a long scrap. If possible, this scrap should be the same type of wood and the same thickness as the stock you want to cut. Scribe a line down the center of the scrap and begin cutting it on the band saw. Adjust the feed direction until the blade cuts straight along the line and you're only making tiny adjustments. Stop cutting and turn off the machine, holding the scrap in place on the worktable. Don't let it move!

4-14 Using the scrap like a straightedge, mark a line on the worktable with a nonpermanent marker, such as a wax pencil or a washable felt-tip pen. You can also use a strip of masking tape. This is the *lead line* for that particular blade as it's presently mounted. When you change blades or readjust this one, you must erase or remove the old lead line and draw a new one.

TRY THIS TRICK

To cut off short parts, make your own miter gauge from a square scrap of wood. Place one edge of the scrap against the fence and hold the stock against the lead edge. Push the scrap forward, keeping it firmly against the fence while feeding the stock into the blade. As long as the fence is aligned with the lead line, this setup will produce accurate crosscuts.

4-15 To adjust a fence parallel to the lead line, measure the distance between the fence and the line in *two* positions — near the infeed end of the fence and again near the out-feed end. Both measurements should be precisely the same.

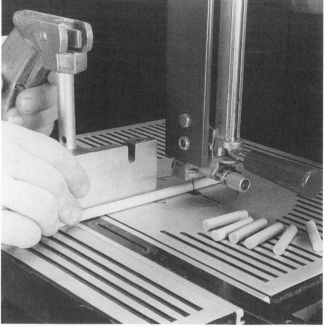

4-16 Because it's so difficult to align the miter gauge slot with the blade lead, you can't easily make guided crosscuts or miters. The miter gauge is normally used only for cutting off narrow stock or small dowels. If the stock is narrow, the cut won't be long enough for the blade to drift very far from the layout line.

AUXILIARY BAND SAW TABLES

The tables on most band saws are small — often too small to safely support the workpiece. When a workpiece is too big for the table, you have two choices — use a saw stand to help support it or attach an auxiliary table to the band saw. Of the two, the auxiliary table offers the best support.

Shown here are two designs for auxiliary tables, one large and the other small. Both surround the worktable on three sides and are attached to some protrusion on the table, such as the rails for a rip fence. If your saw table doesn't have handy protrusions, drill holes in the sides of the worktable and attach ¾-inch-thick, ¾-inch-wide cleats. These cleats should be the same length as the table. Then attach the auxiliary table to the cleats with flathead bolts and wing nuts. This way, the table can be detached easily to change blades.

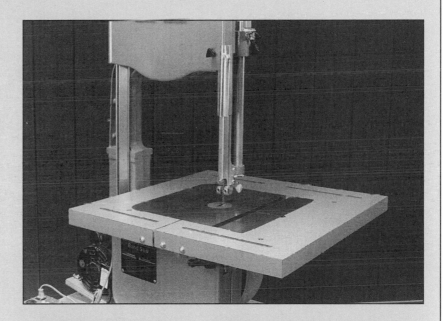

1 **The small auxiliary table** will support most medium-size and some large workpieces. It's designed so it doesn't interfere with the table tilt and other working parts of the band saw. This allows you to leave it attached to the saw, except when changing blades. It has two slots — one on the infeed side and the other on the outfeed — to attach a fence.

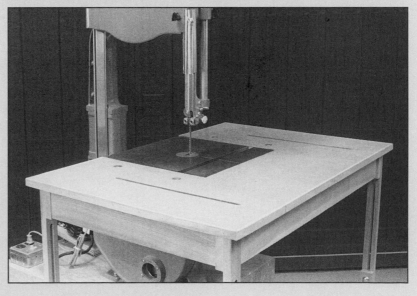

2 **The large auxiliary table will** support large workpieces, including small logs, should you want to use your band saw to do a little lumbering. It has slots for a fence and four adjustable legs for additional support. Unlike the small table, it cannot be tilted.

(continued) ▷

AUXILIARY BAND SAW TABLES — CONTINUED

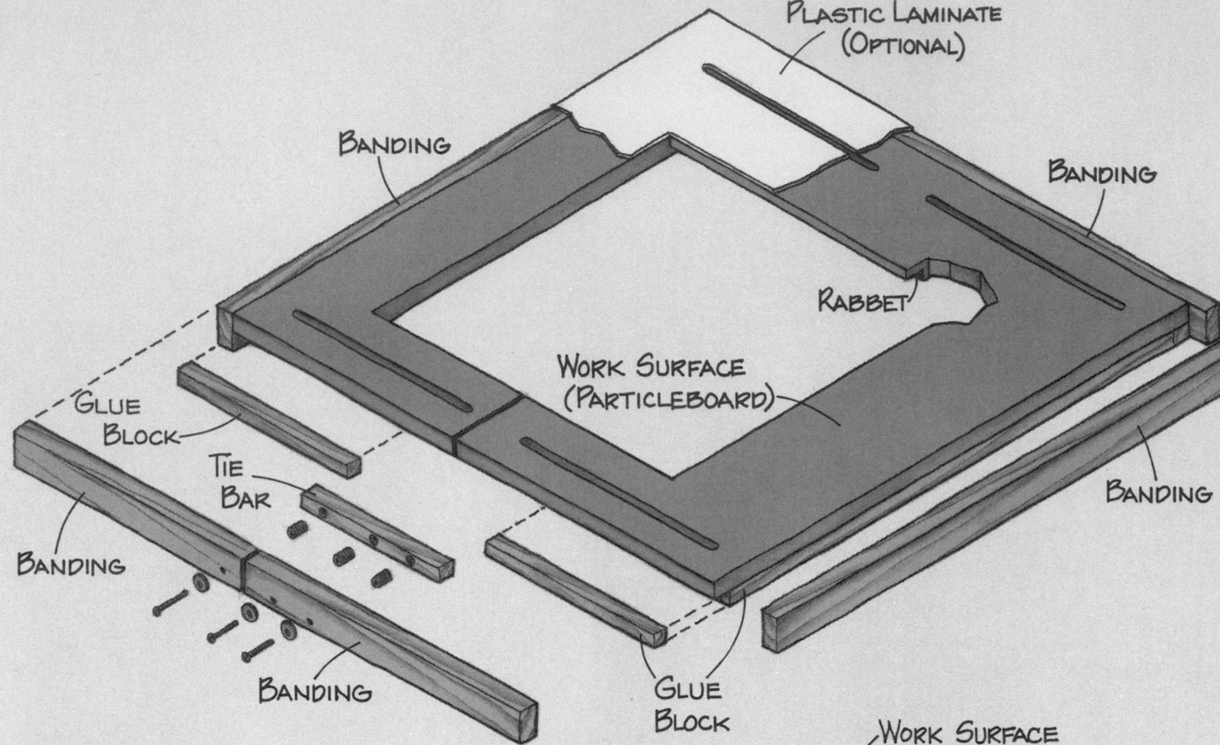

PLASTIC LAMINATE
(OPTIONAL)

BANDING

BANDING

RABBET

WORK SURFACE
(PARTICLEBOARD)

BANDING

GLUE
BLOCK

TIE
BAR

BANDING

BANDING

GLUE
BLOCK

EXPLODED VIEW

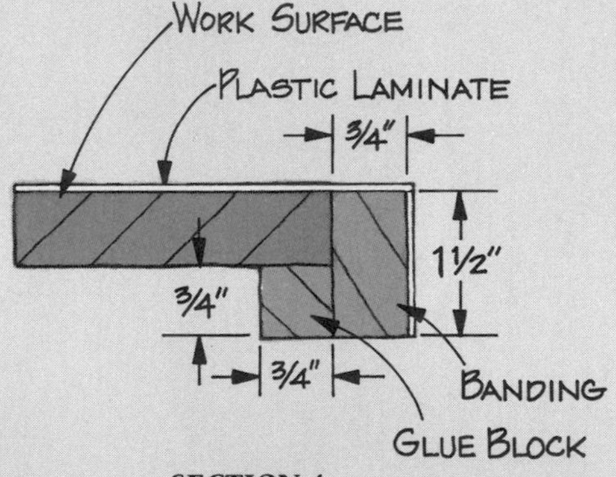

GLUE
BLOCK

¼" x 1¾" R.H.
STOVE BOLT
& WASHER

WORK
SURFACE

¼" x 20
THREADED
INSERT
(3 REQ'D)

GLUE
BLOCK

TIE BAR

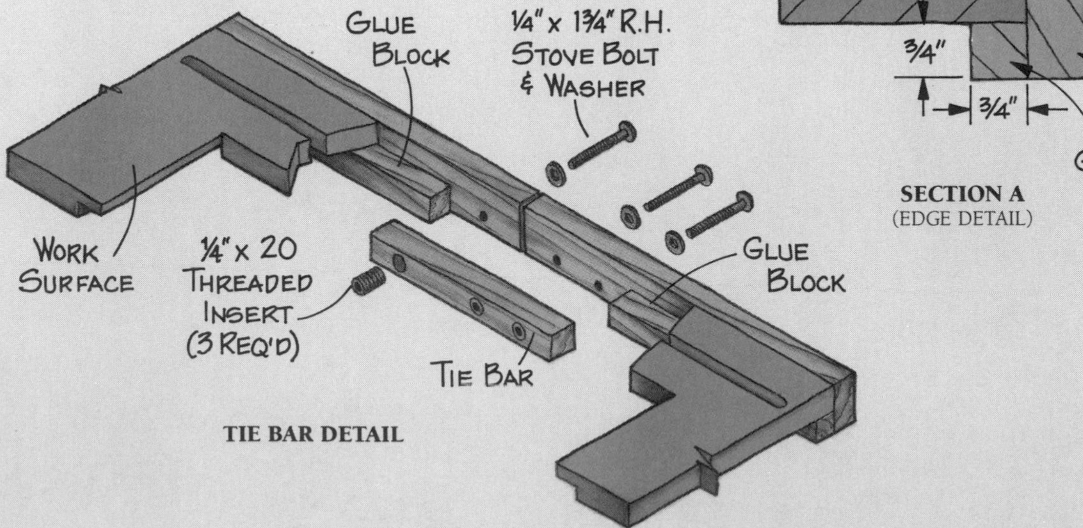

TIE BAR DETAIL

SMALL AUXILIARY TABLE

WORK SURFACE

PLASTIC LAMINATE

¾"

1½"

¾"

¾"

BANDING

GLUE BLOCK

SECTION A
(EDGE DETAIL)

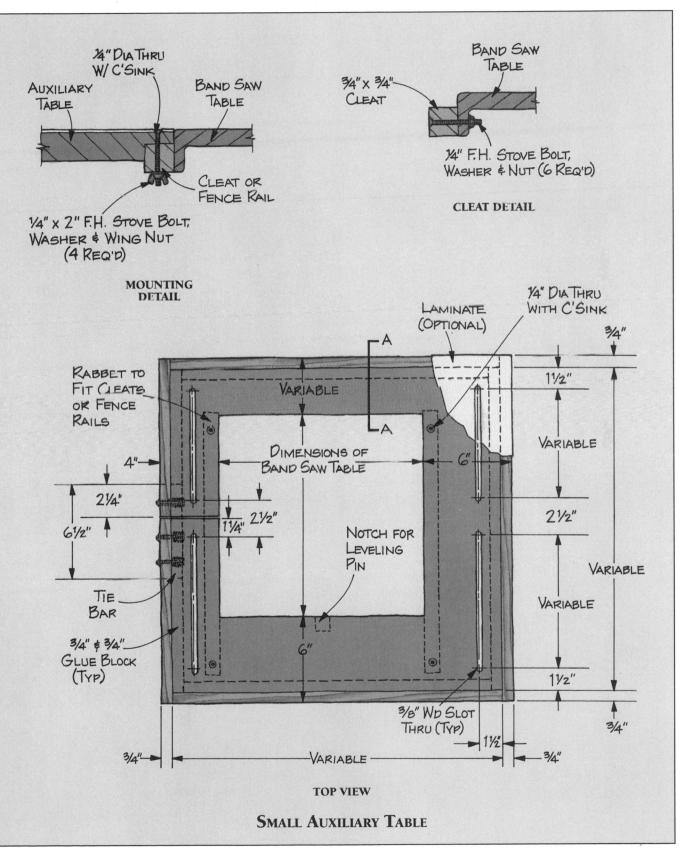

¼" DIA THRU
W/ C'SINK

AUXILIARY
TABLE

BAND SAW
TABLE

CLEAT OR
FENCE RAIL

¼" x 2" F.H. STOVE BOLT,
WASHER & WING NUT
(4 REQ'D)

**MOUNTING
DETAIL**

¾" x ¾"
CLEAT

BAND SAW
TABLE

¼" F.H. STOVE BOLT,
WASHER & NUT (6 REQ'D)

CLEAT DETAIL

LAMINATE
(OPTIONAL)

¼" DIA THRU
WITH C'SINK

¾"

1½"

VARIABLE

RABBET TO
FIT CLEATS
OR FENCE
RAILS

VARIABLE

DIMENSIONS OF
BAND SAW TABLE

6"

4"

2¼"

6½"

1¼" 2½"

NOTCH FOR
LEVELING
PIN

2½"

VARIABLE

VARIABLE

TIE
BAR

6"

¾" & ¾"
GLUE BLOCK
(TYP)

1½"

⅜" WD SLOT
THRU (TYP)

¾"

¾"

VARIABLE

1½"

¾"

TOP VIEW

SMALL AUXILIARY TABLE

(continued) ▷

AUXILIARY BAND SAW TABLES — CONTINUED

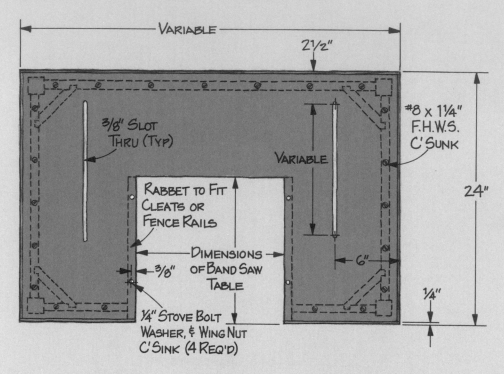

VARIABLE

2½"

3/8" SLOT THRU (TYP)

VARIABLE

#8 x 1¼" F.H.W.S. C'SUNK

24"

RABBET TO FIT CLEATS OR FENCE RAILS

DIMENSIONS OF BAND SAW TABLE

6"

3/8"

¼"

¼" STOVE BOLT WASHER, & WING NUT C'SINK (4 REQ'D)

TOP VIEW

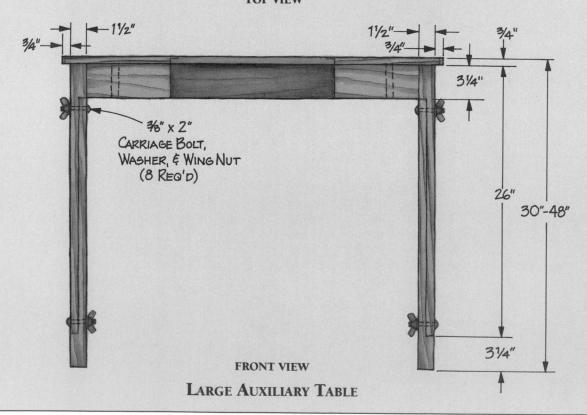

1½"

3/4"

1½"

3/4"

3/4"

3¼"

3/8" x 2" CARRIAGE BOLT, WASHER, & WING NUT (8 REQ'D)

26"

30"-48"

3¼"

FRONT VIEW

LARGE AUXILIARY TABLE

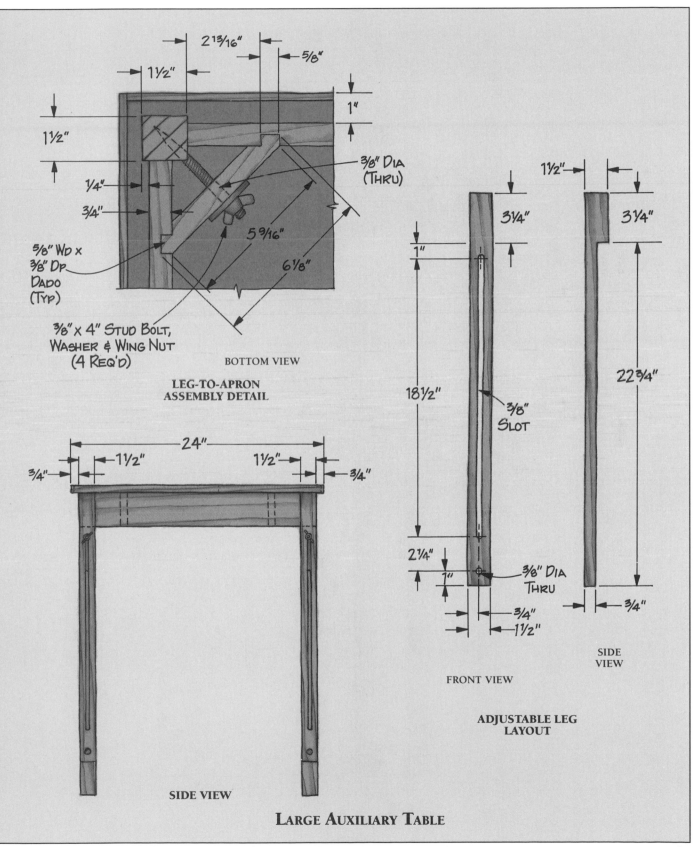

2¹³⁄₁₆"

5⁄8"

1½"

1"

1½"

3⁄8" DIA (THRU)

¼"

3⁄4"

5⁄9⁄16"

5⁄8" WD x 3⁄8" DP DADO (TYP)

6⅛"

3⁄8" x 4" STUD BOLT, WASHER & WING NUT (4 REQ'D)

BOTTOM VIEW

LEG-TO-APRON ASSEMBLY DETAIL

24"

1½"

1½"

3⁄4"

3⁄4"

SIDE VIEW

1½"

3¼"

3¼"

1"

3⁄8" SLOT

18½"

22¾"

2¼"

1"

3⁄8" DIA THRU

3⁄4"

3⁄4"

1½"

FRONT VIEW

SIDE VIEW

ADJUSTABLE LEG LAYOUT

LARGE AUXILIARY TABLE

(continued) ▷

AUXILIARY BAND SAW TABLES — CONTINUED

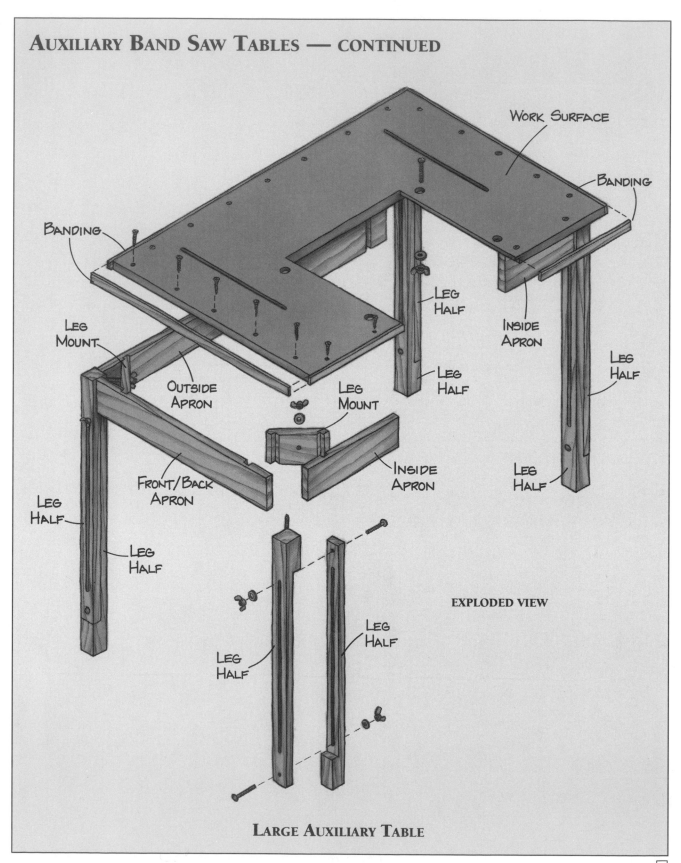

Work Surface

Banding

Banding

Leg Mount

Outside Apron

Leg Mount

Leg Half

Leg Half

Inside Apron

Leg Half

Leg Half

Front/Back Apron

Inside Apron

Leg Half

Leg Half

EXPLODED VIEW

Leg Half

Leg Half

LARGE AUXILIARY TABLE

5

RESAWING

Woodworkers have conceived the first band saw in 1808 as a *resawing* tool for the lucrative new wood art of veneering. Until that time, craftsmen had used a specialized bow saw called a veneer saw to resaw rare, beautiful woods into sheets no more than ⅛ inch thick. They glued these sheets to the surfaces of furniture made from mundane woods, making otherwise ordinary pieces look exotic — and expensive.

Although woodworkers have since found many other uses for the band saw, resawing remains one of its most common applications. You can slice boards with interesting wood grain, then "book match" the parts or arrange them to create vivid designs. And you can resaw lumber for more practical reasons — to save money and materials. If you plane thick stock to make a thin board, you reduce much of the wood to sawdust. But if you resaw the stock and plane the parts, you can make two or more thin boards from a single thick one.

RESAWING TECHNIQUES

CHOOSING THE PROPER METHOD

There are three basic techniques for resawing a board. You can resaw *freehand* with little support and no guidance, following a straight line scribed along an edge. You can use a *pivot* to help support the board while you guide it past the blade. Or, you can use a *fence* to guide *and* support the wood — all you have to do is feed the wood into the blade. *(SEE FIGURE 5-1.)*

Which technique is best? The answer depends on:
- The *width* of the board you must resaw
- The *length* of that board
- *How straight* you must cut the board

For most resawing operations, you must cut through the width of the stock with the board *resting on its edge*. Because the cutting action of the saw holds the stock on the worktable, this is a relatively safe operation as long as the board is stable and has no tendency to tip. The wider the board (in relation to its thickness), the more unstable it becomes. You can resaw narrow boards freehand with no support other than the band saw table itself. But you must use a pivot or a fence to help support wider stock. A good rule of thumb is that you shouldn't freehand stock more than six times as wide as it is thick. If you're resawing ³⁄₄-inch-thick stock, for example, a 4¹⁄₂-inch-wide board is about as wide as you can safely resaw freehand.

The longer the board is, the more difficult it becomes to guide. Because longer boards are more massive,

you must stand farther back from the blade to support them. And the farther you are from the blade, the harder it is to see the scribed line that you're following as you cut. Consequently, it's easier to use a fence to guide boards that are over 4 feet long. And you *must* use a fence if you need to cut a perfectly straight line. When you resaw freehand or with a pivot, the cut will be slightly uneven unless you are very, very steady.

A SAFETY REMINDER

Although resawing is a relatively safe operation on a band saw, it can be dangerous on a table saw or radial arm saw. Most circular saw blades don't have sufficient depth of cut to resaw more than 3-inch-thick stock. You may be tempted to resaw thicker stock by cutting halfway through the board, turning it over, and cutting the remainder. However, this isn't wise. When making a deep cut in which the sawteeth don't clear the top *and* bottom of the kerf, the sawdust may build up in the gullets. As a result, the blade will heat up, expand, and distort. This is hard on the saw and increases the likelihood of kickback.

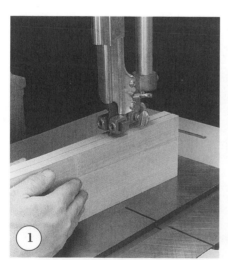

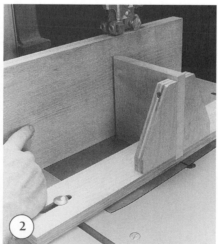

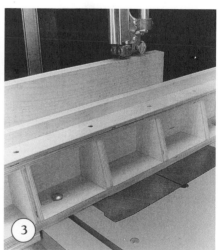

5-1 When resawing *freehand* (1), you must guide the board while it rests on its edge. The board should be stable enough that you can cut it

without horizontal support. If it wants to tip, use a *pivot* (2) to help hold the board upright as you guide it. If the board is too long to maneu-

ver easily or you need to cut a perfectly straight line, use a *fence* (3) to guide *and* support the stock.

PREPARING TO RESAW

No matter what technique you use, there are several details you must attend to before you can resaw safely and accurately.

First, prepare the stock. Joint and plane at least one face of the board — this creates a straight, smooth surface to rest against a fence or pivot. It also gives you a straight surface for the marking gauge to follow, should you need to scribe a line to cut. Most craftsmen prefer to plane *both* faces — this makes the arrises parallel, and it's easier to tell when the blade is drifting too close to one face or the other.

Joint one edge of each board — the edge that the board will rest on. Make sure this edge is square to the planed faces. Then rip the board to the same width all along its length. If the board gets narrower or wider as you cut it, you may have to raise or lower the upper blade guide in the midst of the operation. (SEE FIGURE 5-2.)

Next, prepare the machine. Since you'll be cutting a straight line, mount a wide blade — at least ½ inch wide. A hook-tooth blade works best for resawing, but if the stock isn't very wide or you want a smooth surface, you can also use a skip-tooth or standard-tooth blade. Make certain that the band saw table is square to the blade and that the blade is properly tensioned.

If you're resawing long stock, you must provide additional support for the boards. Attach an auxiliary table to the band saw. Or, place a saw stand at the outfeed side of the table to catch the wood after it's cut.

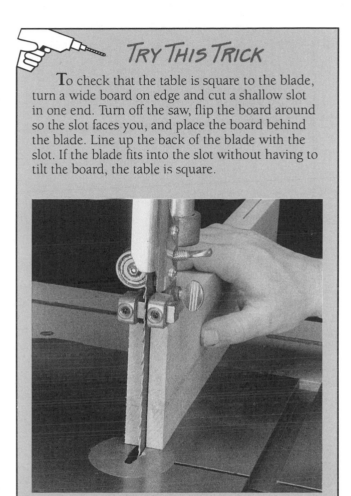

TRY THIS TRICK

To check that the table is square to the blade, turn a wide board on edge and cut a shallow slot in one end. Turn off the saw, flip the board around so the slot faces you, and place the board behind the blade. Line up the back of the blade with the slot. If the blade fits into the slot without having to tilt the board, the table is square.

5-2 Rip each board to a uniform width before resawing it. This lets you adjust the upper blade guide very close to the top edge of the board and leave it there. If the width is uneven, you may have to adjust the upper blade guide as you saw — or set it high enough to accommodate the widest portion of the board. **Note:** When resawing, it's very important to keep the upper blade guide as low as possible. The blade is under a tremendous load, and the closer the upper and lower guides are to one another, the better they will support the blade against the load.

UPPER BLADE GUIDE MUST BE CONSTANTLY READJUSTED TO ACCOMMODATE UNEVEN BOARD

You may also wish to use a saw stand on the infeed side. (SEE FIGURE 5-3.)

Gather up the safety tools you'll need — a push stick or a push shoe to finish the cut, safety glasses to protect your eyes, and a dust mask to protect your nose, throat, and lungs. The dust mask is a good idea even if your band saw is fitted with a dust collector. Resawing creates *lots* of fine sawdust.

Finally, decide how thick you want to resaw the lumber. This depends on several factors:

■ How thick do you want the resawn lumber to be after you've planed the cut surfaces? You must resaw the stock a little thicker than its final dimension to give yourself room to plane it. If you plan to plane both faces of the resawn stock, you'll need to cut it even thicker.

■ How rough is the cut? The deeper the sawmarks left by the blade teeth, the thicker you must resaw the stock. Blades with standard teeth make smoother cuts than skip-tooth or hook-tooth blades; sharp blades cut smoother than dull ones.

■ How trustworthy is your band saw? Resawing demands a lot of performance from a band saw, and some machines are better suited to the task than others. Generally, larger two-wheel saws with powerful motors (1 horsepower or more) do significantly better than smaller tools. If a resawing operation stretches the limits of your band saw, the cut may be uneven. To compensate for this, you must cut the stock thick enough to give yourself room to plane the uneven surface.

So how thick *do* you resaw the stock? If you have a good resawing machine with a reasonably sharp hook-tooth blade, the rule of thumb is to saw the stock at least 1/16 inch thicker than its final dimension *for each surface that you intend to plane.* For example, if you need 1/4-inch-thick stock and will only plane one face, saw the stock 5/16 inch thick or thicker. If both faces of the resawn stock will require planing, slice the wood at least 3/8 inch thick.

RESAWING FREEHAND

To resaw a board freehand, first scribe a line along the top edge. (SEE FIGURE 5-4.) You'll follow this line as you cut. Place the board on the saw and check that the face is parallel to the blade. (SEE FIGURE 5-5.) If the setup is satisfactory, feed the wood into the blade slowly and gently. Try to find the blade lead, then feed as straight as possible. Go slowly enough that you can adjust the feed direction the moment the blade drifts from the line. When it does drift, make small, gradual changes in the feed direction. This will give you the smoothest possible freehand cut.

5-3 To provide extra support when resawing long stock, place a saw stand to catch the wood as it leaves the band saw table. You can purchase these saw stands from many mail-order woodworking suppliers, or you can make your own. The stand shown here was made from a rolling pin and construction lumber.

5-4 When you're sawing freehand or with a pivot, use a marking gauge to scribe a line in the top edge of the stock. This will give you an accurate line to follow as you cut. **Note:** Although it's not necessary, scribing a line is also useful when resawing with a fence. Using the line as a reference, you can tell if the blade is drifting or wandering as it cuts.

RESAWING WITH A PIVOT

To resaw with a pivot, you must rest the face of the board against a fixed block as you feed the wood past the blade. This block is secured to the band saw table, and the distance between the pivot block and the blade determines how thick the stock will be after it's resawn.

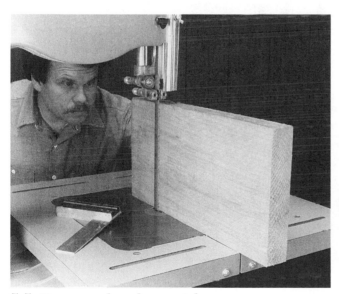

5-5 To resaw a board accurately, the table must be perpendicular to the blade, and the face of the board must rest perpendicular to the table. This will make the blade parallel to the face of the board. If the blade and the board are *not* parallel, the resawn stock will not be a uniform thickness — it will be wider at one edge than the other. To check that the blade and board are parallel, stand the board on edge on the work-table. Slide the face of the board against the blade and sight between them. If they diverge at the top or the bottom, either the table tilt is improperly adjusted or the edge has not been jointed properly.

The guiding edge of the pivot block must be *pointed*, not round, to provide a true "pivot point." When resawn stock rotates around a rounded edge, it moves sideways in relation to the blade. This movement is very slight, but it's enough to distort the blade. Any such distortion will detract from the quality of the cut. When you mount the pivot on your machine, check that the pointed edge is parallel to the blade. (*See Figure 5-6.*) Position the point dead even with the blade teeth, as you look at the band saw from the side. (*See Figure 5-7.*)

Once you've positioned the pivot, proceed in much the same manner as you would with a freehand cut. Scribe a guiding line on the top edge of the stock, then feed the stock into the blade, keeping one face against the pivot point. If the blade drifts from the line, adjust the feed direction, but always keep the stock firmly against the pivot.

Some craftsmen prefer to use two pivots, one on either side of the board. (*See Figure 5-8.*) Opposing pivots keep the wood upright with no need to apply sideways pressure. Resawing wide boards will be easier, although it takes longer to set up.

5-6 To resaw accurately with a pivot, the pivot must be parallel to the blade. Check this by placing the pivot block on the band saw work-table and sliding the pointed edge against the body of the blade. The point should contact the blade all along its length. If it doesn't, adjust the table tilt as necessary.

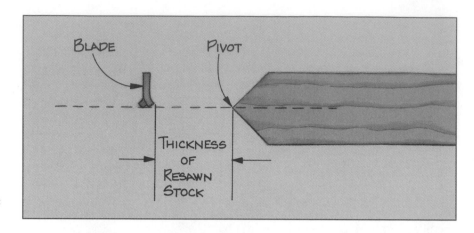

5-7 Attach the pivot block to your band saw so the point is even with the teeth. (Imagine a line drawn perpendicular to the body of the blade, extending out from the teeth. Position the pivot along this line.) The distance from the pivot point to the blade will determine the thickness of the resawn stock.

5-8 You can resaw with one pivot or two, depending on your preference. It's easier to set up one pivot, but you must apply constant sideways pressure against the wood as you cut it to keep the stock against the pivot. A second pivot on the opposite side of the blade eliminates the need for this pressure, keeps the stock upright, and leaves you free to concentrate on following the cut line. To use two pivots, *both* sides of the stock must be planed, and the distance between the pivots must be precisely the same as the thickness of the planed stock.

RESAWING WITH A FENCE

The preferred arrangement for resawing wide boards is with a fence. A fence supports *and* guides the stock as you cut, so all you have to worry about is the feed rate.

A fence, however, requires more care than a pivot when setting it up. Not only must the fence be the proper distance from the blade, it must be parallel to both the blade *and the lead line* for that blade. (SEE FIGURES 5-9 AND 5-10.) If any of these adjustments are off, the cut will be uneven or inaccurate.

You may wish to use one or more featherboards to hold the wood against the fence as you feed it past the blade. (SEE FIGURE 5-11.) Although featherboards are useful no matter what the size of the stock, they are a necessity when resawing long boards. Because you must stand farther away from the blade when resawing a long board, it's difficult to hold the board flat against the fence without featherboards.

Feed the wood into the blade slowly, giving the blade plenty of time to clear the sawdust from the cut. The wider the board, the more sawdust there is to clear, and the more slowly you must feed the stock. Be especially careful as you come to the end of the cut and the blade emerges from the stock. When you cut wide stock, there's a lot of blade exposed at that moment.

FOR BEST RESULTS

When using a resawing pivot or fence, the jig should be somewhat shorter than the width of the stock you want to resaw. If the pivot or fence is taller than the stock is wide, you won't be able to adjust the upper blade guide properly. The blade will be partly exposed and it won't be supported in the cut as well as it could be.

5-9 To set up a fence for resawing, first find the lead for the blade you intend to use, as described in "Making Guided Cuts" on page 44. Rest the fence on the band saw table so the distance from the blade to the fence is equal to the thickness you wish to cut. Measure the distance from the fence to the lead line near the infeed and outfeed edges of the table — these measurements should be the same. When the fence is properly positioned, clamp it to the table and check the measurements again.

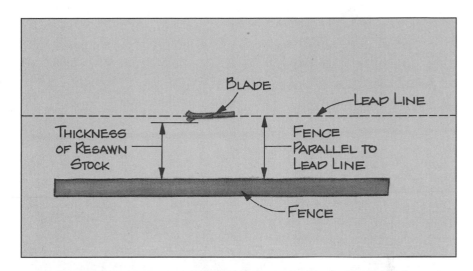

5-10 For the thickness of the stock to be even, the fence must be parallel to the blade. Measure the distance from the blade to the fence near the top and bottom of the fence — these two measurements should be the same. If they aren't, adjust the table tilt to make them so. This should be done *after* the fence is clamped to the table. Unless the fence is perfectly mated to the table, it may distort slightly when clamped in place. By making the tilt adjustment after the fence is secured, you can compensate for any distortion.

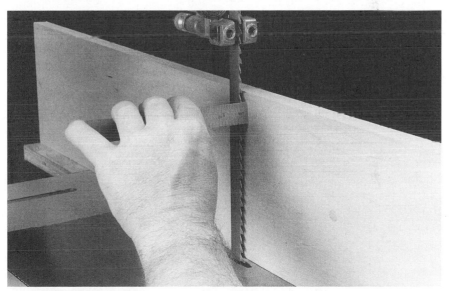

5-11 Featherboards help hold the wood flat against the fence while you feed it past the blade. Position these safety tools to press against the wood *before* it gets to the blade. If they press against the stock at the blade or behind it, the featherboards will pinch the sides of the kerf together. This, in turn, increases the drag on the blade. It will bog down, heat up, and may break. (For plans and instructions on how to make the featherboard jig shown, see "Resawing Aids" on page 61.)

RESAWING TROUBLESHOOTING

Because resawing makes tremendous demands on the band saw, problems can occur if the tool, the pivot, or the fence isn't properly set up. To catch these problems before you cut good stock, *always cut a test piece to check the setup.* This test piece should be at least 12 inches long and as wide as the stock you intend to resaw.

As you cut the test piece, look and listen for problems. Does the machine bog down easily? Does the blade wander back and forth? Does it pull to one side? After you finish the cut, inspect the piece for more problems. Is the cut piece the proper thickness? Is the thickness even across the width of the board? Most of these problems can be corrected simply. If the machine slows down, you may need to reduce the feed rate or change to a sharper blade. Other problems can often be solved by checking the alignment of the band saw and the position of the pivot or fence.

The cut surfaces also may be cupped or extremely rough and uneven. Cupping occurs when the blade bows or "bellies" in the wood, cutting a curve through the width of the stock. This is caused by uneven forces on the blade. The feed pressure against the teeth compresses the leading edge of the blade while stretching out the trailing edge. The blade wants to buckle and, when it does, it usually follows the path of least resistance, cupping in the same direction as the annual rings in the wood. (*SEE FIGURE 5-12.*)

Cupping can be corrected in several different ways. You can reduce the feed pressure or use either a wider blade or a sharper one. But the most common way is to increase the blade tension. Resawing operations often require that the blade be tensioned higher than normal. Remember, you can safely increase the tension one notch on the tension scale.

A SAFETY REMINDER

Don't adjust the tension so high that it collapses the tension spring. (The coils must not close completely.) When the coils are open, this spring acts as a shock absorber for wheels that may be slightly out of round. Once the coils of the spring are completely compressed, this buffer is gone. Both the blade and the band saw may be damaged.

A rough, uneven cut, often called *washboarding,* is caused by excessive vibration. Because the blade is under tension, it acts like a big guitar string — it picks up small vibrations in the moving parts and amplifies them. As it vibrates, the blade slaps back and forth in the kerf so the teeth scrape one side of the cut and then the other. This makes the cut surface look like an old washboard. (*SEE FIGURE 5-13.*)

5-12 It's always wise to check your setup by cutting a scrap before resawing good wood — this will reveal any problems before it's too late to correct them. One of the most common problems is *cupping* — the blade twists and bows in the cut, leaving the cut surfaces curved rather than flat. Usually, this can be corrected by applying more tension to the blade — you can safely tighten the blade one step above its indicated level on the tension scale, as long as you do not collapse the tension spring. (For example, you can safely tighten a ½-inch-wide blade to the ¾-inch level.) If the cupping persists, use a wider blade or feed the wood more slowly.

When cutting thin stock, washboarding is minimized because the blade guides are closer together. The portion of the blade that vibrates in the cut is shorter and the side-to-side movement is smaller. But when you're resawing, the guides are farther apart and this movement is greater. If your band saw vibrates excessively, it may be impossible to get a smooth cut no matter how well you've aligned and adjusted the machine.

If you're plagued by washboarding, the cause may be damaged or worn tires, out-of-round wheels, wobbling pulleys, or a cracked or poorly made drive belt. Carefully track down the source of the problem and either replace the bad part or have it fixed.

5-13 If the band saw blade vibrates excessively as it cuts, the cut surface will be rough and uneven. The wider the resawn board and the further apart the blade guides, the worse this *washboarding* becomes. To get a smooth cut, you must track down the source of the vibration and fix it.

RESAWING AIDS

To resaw wide boards on your band saw, you will have to make your own resawing jigs — pivots, fences, and featherboards. Pivots and featherboards aren't offered commercially as band saw accessories. And manufactured fences are too short to be used for resawing — you must make a tall fence extension. Fortunately, all of these jigs are easy to make.

To make a *Resawing Pivot,* cut a double bevel in one end of a hardwood board to create a pivot point. Attach this pivot board to a base so you can clamp it to the band saw table. If you wish, cut slots in the base (as shown in the *Top View*) so you can secure the assembly to the *Small Auxiliary Table* on page 47. Brace the pivot so it remains upright. When it's clamped to the band saw, the angle between the pivot and the table must be exactly 90 degrees.

To make a *Resawing Fence,* cut a wide board about 12 inches longer than the depth of your band saw table (the distance from the infeed side to the outfeed side). Attach horizontal and vertical braces to the fence board to keep it straight and flat. Then attach the assembly to a base so it can be clamped to the band saw table. (Like the pivot base, the fence base is designed so that it can be attached to the auxiliary table.) Add ledges to both the infeed and outfeed ends of the fence for additional support.

The *Adjustable Featherboards* are made from two ordinary featherboards attached to a stand. To make this stand, mount two 3/4-inch-diameter dowels in a base. Drill 3/4-inch-diameter holes in each featherboard to fit over the dowels. Cut a slot through both holes and drill a 1/4-inch-diameter hole perpendicular to the slot. Insert a carriage bolt with a washer and wing nut through the 1/4-inch-diameter hole. This will act as a clamp to secure the featherboard to the dowels.

(continued) ▷

RESAWING AIDS — CONTINUED

1 **All of the resawing jigs —** pivot, fence, and featherboards — are designed to be bolted to an auxiliary table *or* clamped to the band saw table. To make clamping easier, cut small wooden blocks to fill the spaces between the ribs on the underside of the table. Plane these blocks precisely as wide as the ribs — you don't want them to protrude from beneath the table. Secure them with contact cement.

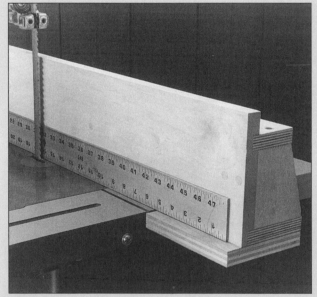

2 **The ledges on the *Resawing Fence*** must be flush with the table surface when the fence is clamped to the band saw. The ledges provide additional support on both the infeed and outfeed side of the table. This, in turn, makes it easier to resaw long stock.

3 **To hold a wide board securely** against the fence, the *Adjustable Featherboards* should press against the stock near the top *and* bottom edges. To raise or lower each featherboard, loosen the wing nut that clamps it to the dowels, raise or lower it to the desired position, and tighten the nut.

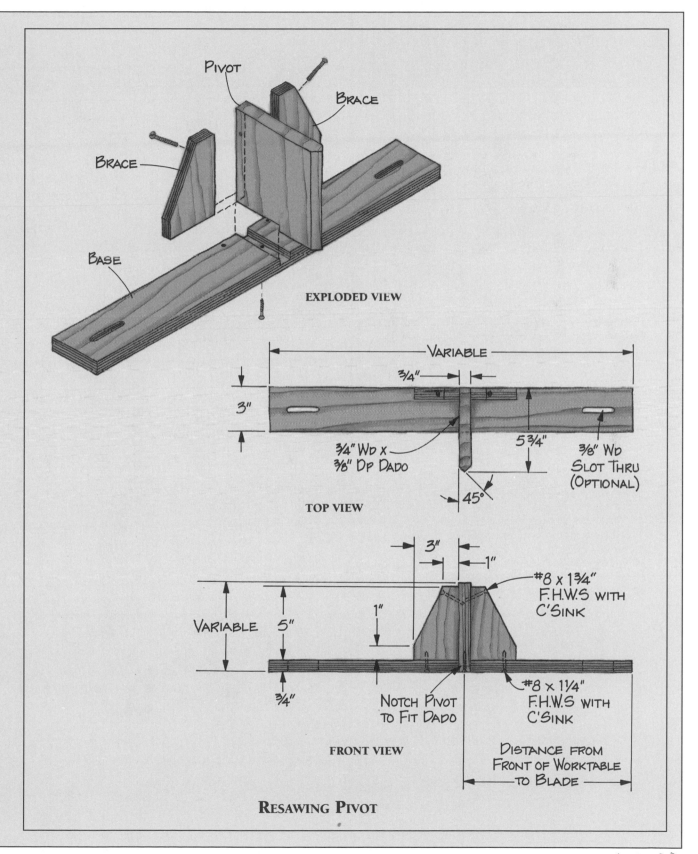

PIVOT

BRACE

BRACE

BASE

EXPLODED VIEW

VARIABLE

3/4"

3"

3/4" WD X
3/8" DP DADO

45°

5 3/4"

3/8" WD
SLOT THRU
(OPTIONAL)

TOP VIEW

3"

1"

#8 x 1¾"
F.H.W.S WITH
C'SINK

VARIABLE

5"

1"

3/4"

NOTCH PIVOT
TO FIT DADO

#8 x 1¼"
F.H.W.S WITH
C'SINK

DISTANCE FROM
FRONT OF WORKTABLE
TO BLADE

FRONT VIEW

RESAWING PIVOT

(continued) ▷

RESAWING AIDS — CONTINUED

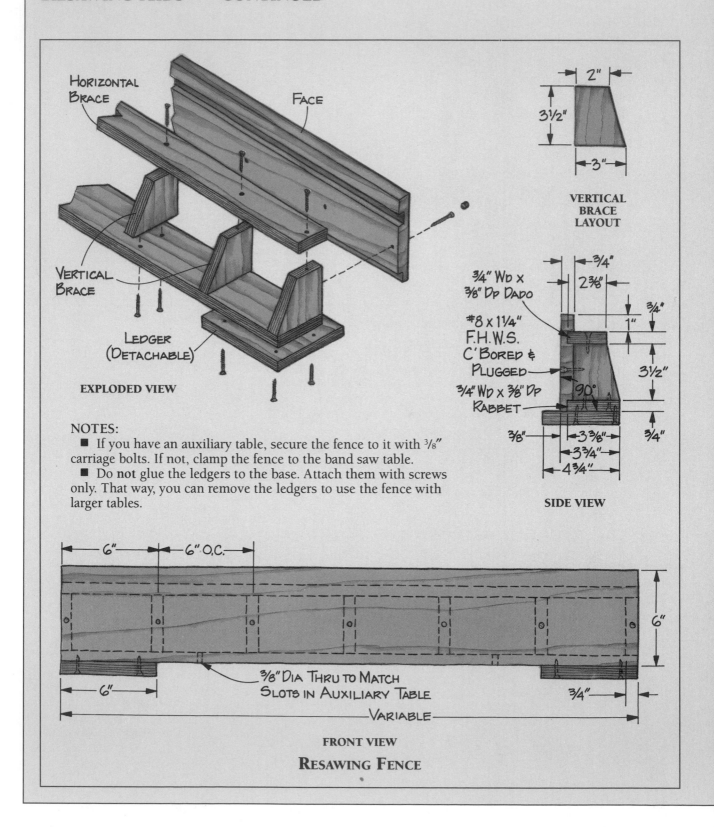

HORIZONTAL BRACE

FACE

VERTICAL BRACE

LEDGER (DETACHABLE)

EXPLODED VIEW

2"

3½"

3"

VERTICAL BRACE LAYOUT

¾" WD X ⅜" DP DADO

#8 x 1¼" F.H.W.S. C'BORED & PLUGGED

¾" WD X ⅜" DP RABBET

¾"

2⅜"

¾"

1"

3½"

¾"

90°

⅜"   3⅜"
          3¾"
         4¾"

SIDE VIEW

NOTES:
■ If you have an auxiliary table, secure the fence to it with ⅜" carriage bolts. If not, clamp the fence to the band saw table.
■ Do **not** glue the ledgers to the base. Attach them with screws only. That way, you can remove the ledgers to use the fence with larger tables.

6"   6" O.C.

6"

6"

6"   ⅜" DIA THRU TO MATCH SLOTS IN AUXILIARY TABLE   ¾"

VARIABLE

FRONT VIEW
RESAWING FENCE

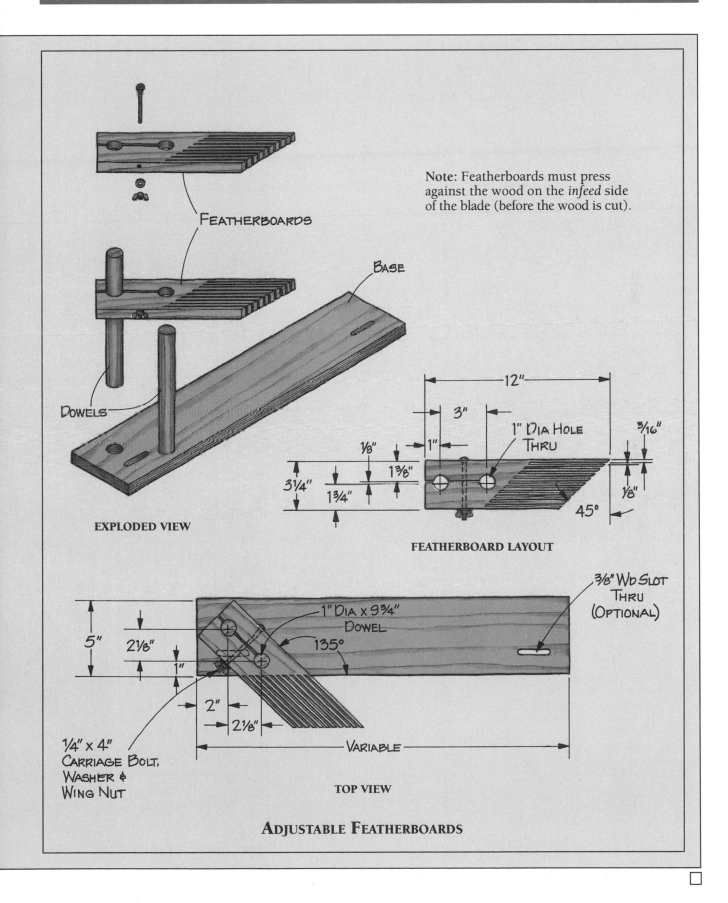

FEATHERBOARDS

BASE

DOWELS

Note: Featherboards must press against the wood on the *infeed* side of the blade (before the wood is cut).

EXPLODED VIEW

12"

3"

1"

1" DIA HOLE THRU

3/16"

1/8"

1 3/8"

3 1/4"

1 3/4"

1/8"

45°

FEATHERBOARD LAYOUT

3/8" WD SLOT THRU (OPTIONAL)

5"

2 1/8"

1"

1" DIA x 9 3/4" DOWEL

135°

2"

2 1/8"

1/4" x 4" CARRIAGE BOLT, WASHER & WING NUT

VARIABLE

TOP VIEW

ADJUSTABLE FEATHERBOARDS

RESAWING APPLICATIONS

REASONS TO RESAW

Although most woodworkers think of resawing as making thin boards out of thick ones, there are many other uses for resawing techniques. For instance, veneer cutters have found that they can split a board with a striking grain pattern, then *book match* the halves (use them side by side) so the grain pattern in each half is a mirror image of the other. (*SEE FIGURE 5-14.*) Using a variation of this technique, you can glue up woods of contrasting colors, slice the glued-up stock, then arrange the pieces to make herringbone, checkerboard, and other geometric patterns. (*SEE FIGURE 5-15.*) Or, you can resaw duplicates of intricately shaped parts — cut a shape in a thick board, then resaw it into several thin ones. Each slice will have exactly the same profile. (*SEE FIGURE 5-16.*)

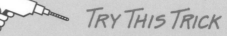

TRY THIS TRICK

Instead of resawing a shape to make duplicate parts, stack several thin boards face to face, holding the stack together with double-faced carpet tape, and cut a shape in the entire stack. When you remove the tape, each piece will be exactly the same shape. This is sometimes referred to as *pad sawing*.

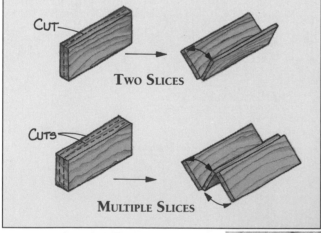

TWO SLICES

MULTIPLE SLICES

5-14 Burls and other figured woods often have beautiful "wild" grain — the stock has no grain direction; the wood fibers twist and turn randomly. You can use this random wood grain to create intricate patterns by resawing the stock and *book matching* the slices. Carefully mark the edges before you cut the stock, then arrange the slices edge to edge, always joining the edge on one slice to the *same* edge on the adjoining slice.

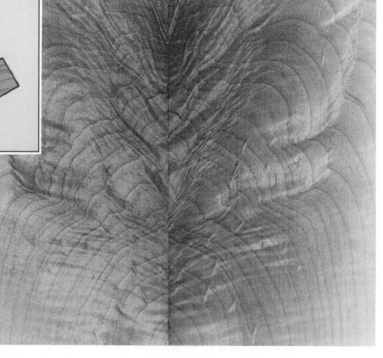

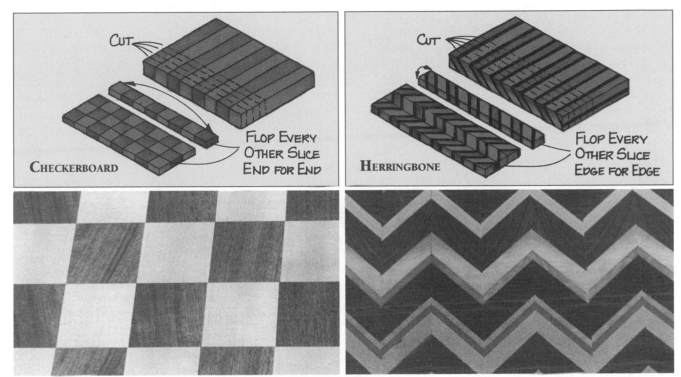

5-15 You can create many striking geometric patterns with ordinary woods. Glue up scraps of contrasting colors, resaw the glued-up stock *across* the glue joints to make multi-colored strips, and arrange these strips edge to edge. The resulting patterns depend on the colors you choose, how you join the scraps, how you slice the glued-up stock, and how you arrange the slices. The boards in this herringbone cutting board were glued up at an angle, then the resawn strips were turned edge for edge. The strips in the checkerboard were sliced square to the glue joints, then every other strip was flopped end for end.

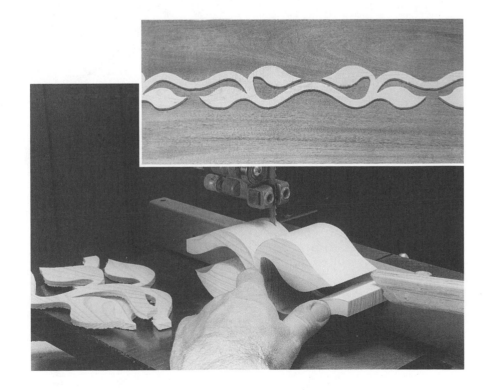

5-16 To make this applied molding, first cut the profile of a grape vine and leaves in a wide board. Then resaw the board into thin strips. When laid end to end, the vine shape will repeat over and over again.

Resawing comes in handy when you run across limbs and logs that are too small to take to a sawyer, but too valuable to turn into firewood. With an auxiliary table and a simple log carriage, you can turn your band saw into a small lumber mill. (*See Figure 5-17.*) Technically, this operation is *sawing* rather than resawing, since the wood hasn't been sawn yet, but the techniques are precisely the same.

You can also resaw stock to make boxes, cases, and small chests from single blocks of wood. Instead of slicing a block in one direction only, resaw from several different directions, removing slices from different surfaces. Plan your cuts carefully so you can glue the slices back together to make an enclosed box. (*See Figure 5-18.*) If you carefully match the wood grain as you glue the resawn pieces back together, the completed box will look as if it were carved from a single, uncut piece of wood. **Note:** Use a standard or skip-tooth blade for a smooth cut.

5-17 With a special jig to hold the stock, you can use resawing techniques to turn small logs and limbs into usable lumber. For plans and instructions, refer to "Band Saw Lumbering" on the facing page.

5-18 This entire set of nesting boxes was cut from a single block of wood by resawing the parts, then gluing them back together again. The parts for each box were cut in this order: first the *lid,* then the *front* and *back,* then the *sides,* then the *lid* and *bottom.* The leftover stock became the block for the next smaller box. For complete instructions, see "Nesting Boxes" on page 111.

BAND SAW LUMBERING

Have you ever run across cherry or walnut logs in your wood pile and thought they would make better lumber than firewood? Or have you ever cut down a small fruit tree and wished you knew of a sawmill that would take small logs? If you have a band saw, you can *make* that sawmill. By using the *Large Auxiliary Table* (shown on page 47) and a simple *Log Carriage* to hold and guide the wood, you can cut small logs and tree limbs up to 4 feet long into usable lumber. The diameter of the stock you can cut with this setup depends on the depth of cut of your machine.

The log carriage consists of a 12-inch-wide base and an adjustable fence. The fence moves in or out to accommodate small and large logs. These logs are fastened to the fence with lag screws to keep them from rolling or sliding around on the carriage.

The entire carriage slides on the auxiliary table as you feed the log past the blade. To keep the carriage moving in a straight line, it's guided by an *Adjustable Guide Rail* near the outside edge of the table. The rail can be angled right or left to correct for blade lead.

To saw a small log into lumber, mount a wide hook-tooth blade on your band saw, attach the auxiliary table to the machine, and secure the *Adjustable Guide Rail* to the table. Find the lead for the blade, then adjust the rail parallel to the lead line and 12 inches away from the blade. (The distance between the blade and the rail must be equal to the width of the carriage.) Depending on the length and size of the logs, you may also have to place saw stands near the infeed and outfeed ends of the table.

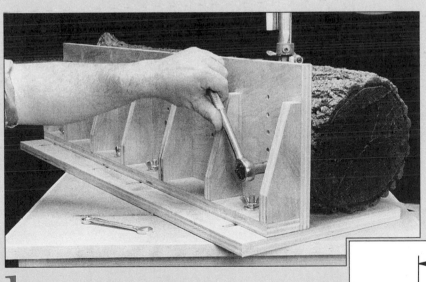

1 Place the log on the carriage and adjust the fence to hold the log so approximately 15 percent of the log's diameter overhangs the inside edge of the base. This is the portion of the log that will be sliced off. For example, if you're sawing a log 8 inches in diameter, you want to cut off a slab about 1¼ inches thick, so the log should overhang 1¼ inches. Fasten the log to the fence with at least two lag screws.

(continued) ▷

BAND SAW LUMBERING — CONTINUED

2 **Place the carriage on the** table so the outside edge of the base rests against the rail. Turn on the saw and slide the carriage forward, feeding the log into the blade. As you cut, keep the carriage firmly against the rail. This will create one flat side on the log.

3 **Turn the log 90 degrees** on the carriage so the flat side faces down. Again, position the log to cut away about 15 percent of the diameter. Secure the log and cut a second slab, creating another flat side square to the first. Repeat this process for each of the logs that you want to saw up. **Note:** If you wish, you can make more than two flat sides on each log. If you cut away four slabs, reducing the log's diameter by 15 percent with each cut, you'll saw the log into a square beam.

4 **Once you've cut at least two** flat sides on all the logs, set the log carriage aside and remove the adjustable rail from the auxiliary table. Attach a fence in its place. (Use the fence from "Resawing Aids" on page 61.) Adjust this fence as you would a resawing fence and saw the logs into boards. As you cut, keep one flat side against the table and another against the fence.

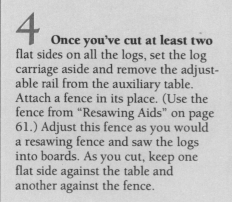

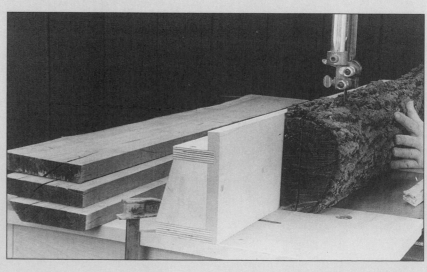

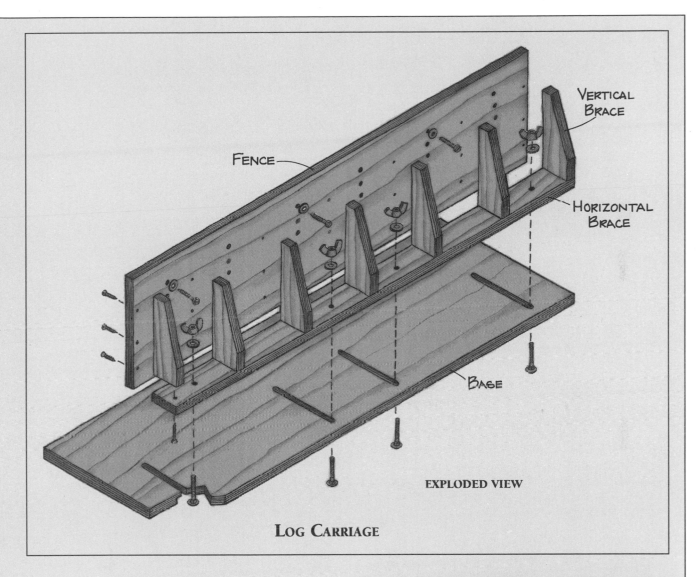

FENCE

VERTICAL
BRACE

HORIZONTAL
BRACE

BASE

EXPLODED VIEW

LOG CARRIAGE

5 **Stack the wood in an out-**of-the-way place to dry. Arrange it so there's a small space between the edges of each board and any adjacent boards, and place spacers (called *stickers*) in between each layer of boards. This will allow the air to circulate over all the surfaces of all the boards. Let the boards sit until their moisture content stabilizes. The rule of thumb is to dry them one year per inch of thickness.

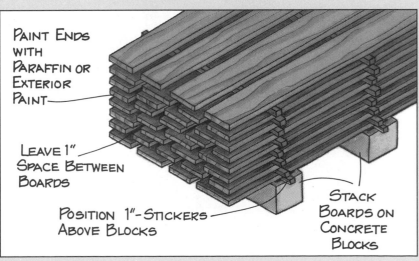

PAINT ENDS
WITH
PARAFFIN OR
EXTERIOR
PAINT

LEAVE 1"
SPACE BETWEEN
BOARDS

POSITION 1"-STICKERS
ABOVE BLOCKS

STACK
BOARDS ON
CONCRETE
BLOCKS

(continued) ▷

BAND SAW LUMBERING — CONTINUED

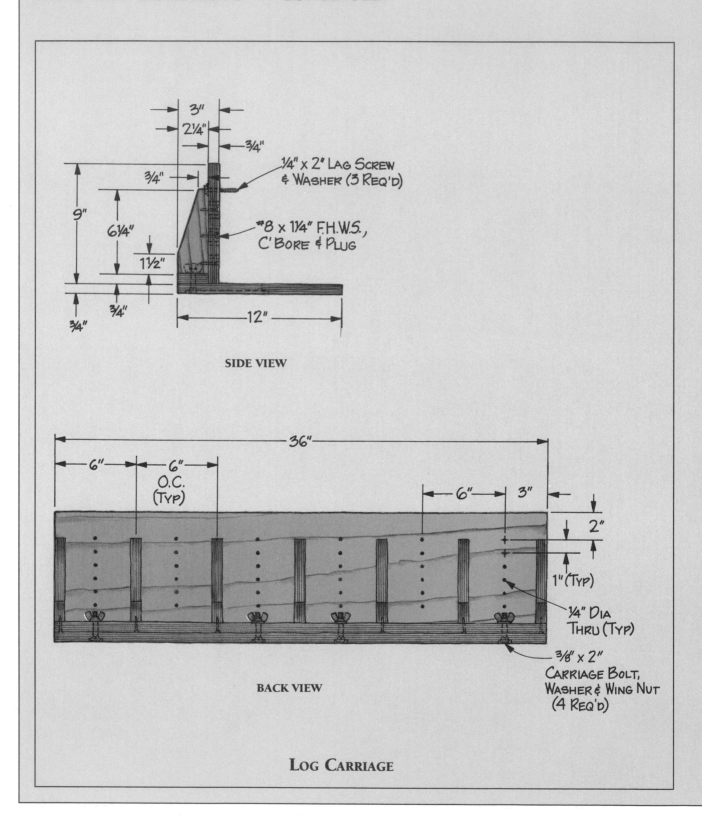

¼" x 2" Lag Screw
& Washer (3 Req'd)

#8 x 1¼" F.H.W.S.,
C'Bore & Plug

3"

2¼"

¾"

¾"

9"

6¼"

1½"

¾"

¾"

12"

SIDE VIEW

36"

6"

6"
O.C.
(Typ)

6"

3"

2"

1" (Typ)

¼" Dia
Thru (Typ)

⅜" x 2"
Carriage Bolt,
Washer & Wing Nut
(4 Req'd)

BACK VIEW

LOG CARRIAGE

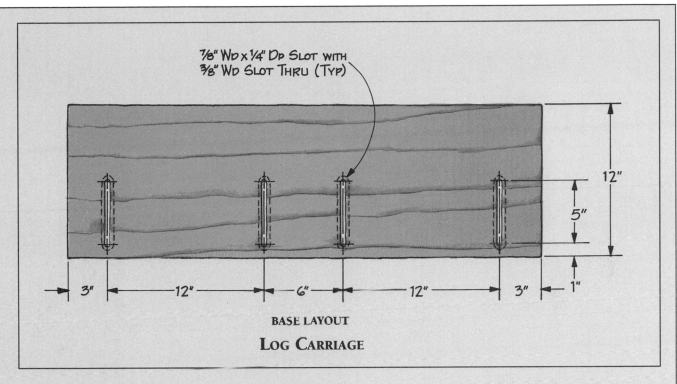

7/8" Wᴅ x 1/4" Dᴘ Sʟoᴛ ᴡɪᴛʜ
3/8" Wᴅ Sʟoᴛ Tʜʀᴜ (Tʏᴘ)

12"

5"

1"

3"   12"   6"   12"   3"

BASE LAYOUT
Log Carriage

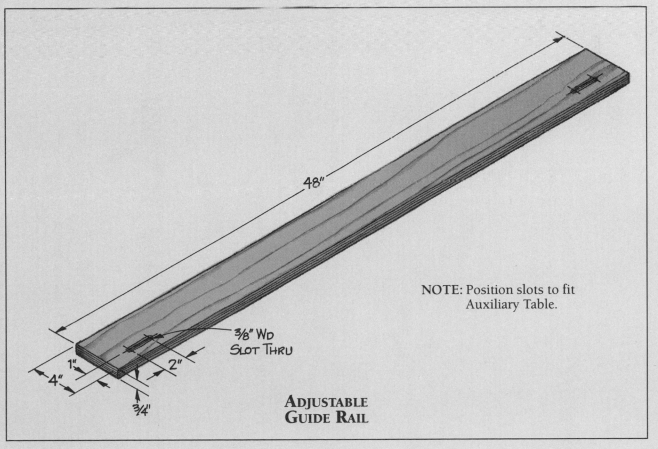

48"

NOTE: Position slots to fit
Auxiliary Table.

3/8" Wᴅ
Sʟoᴛ Tʜʀᴜ

1"

2"

4"

3/4"

Adjustable
Guide Rail

6

SPECIAL BAND SAW TECHNIQUES

Most band saw operations are simple tasks, such as cutting two-dimensional patterns or resawing thick stock into thin boards. But the band saw is an extraordinarily capable machine and can do much more when you need it to.

For example, you can make *three-dimensional* shapes with the band saw by cutting two patterns in different surfaces of the same board. Depending on the patterns used and how you combine them, you can create elegant cabriole (S-shape) legs, realistic and abstract figures, and other sculptured shapes.

With a circle-cutting jig, you can saw perfect circles of almost any size on the band saw. Still other jigs and special techniques make it possible to cut dovetail and finger joints.

MAKING COMPOUND CUTS

BASIC TECHNIQUE

You create three-dimensional shapes on the band saw by cutting two or more patterns in the same workpiece, making each cut in a different plane. This technique is often referred to as *compound cutting*. The basic technique involves just three steps (*SEE FIGURES 6-1 THROUGH 6-3*):

1. Cut a pattern in one surface of a rectangular workpiece.
2. Tape the waste to the workpiece to make the stock rectangular again.
3. Turn the stock 90 degrees and cut another pattern in a second surface.

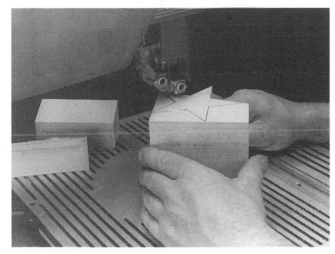

6-1 To make a compound cut, first trace a two-dimensional pattern on one surface of a rectangular workpiece. It doesn't matter which surface you begin with — it can be a face, edge, or end. Cut the pattern on the band saw, carefully saving each piece of waste.

6-2 Tape the waste back to the workpiece to make the block rectangular again. Turn the block 90 degrees so an uncut surface faces up. Trace a second pattern on this new surface.

6-3 Cut the second pattern in the workpiece. After removing all the waste, you'll find that the intersecting cuts create a three-dimensional shape.

When you remove the tape and discard the waste, the workpiece will display *compound contours* — two-dimensional shapes that intersect at right angles. In doing so, they combine to make one three-dimensional shape. This shape becomes increasingly complex as you cut patterns in additional surfaces. (*See Figure 6-4.*) **Note:** When cutting a series of *different* patterns, it often helps to cut the simplest patterns first. The chunks of waste will usually be larger and easier to tape in place.

TRY THIS TRICK

Use double-faced carpet tape to secure the waste to the workpiece when making compound cuts. This not only keeps the waste from shifting (as sometimes happens when you use other types of tape), but also helps to fill in the void created by the saw kerf. The reassembled stock is closer to its original dimensions than it might be otherwise.

6-4 Usually, you make just two cuts 90 degrees apart when compound cutting. However, you can cut as many different patterns in as many different planes as you want. Each cut will make the shape increasingly complex. These three shapes show what happens when you cut a simple circle in one, two, and three planes.

BAND SAW SCULPTURE

Compound cutting is often used to prepare blocks for woodcarving. Draw two views of the subject from two different angles, 90 degrees apart. Trace one view on the carving block, cut the outside shape, and tape the waste back to the block. Then trace and cut the second view in a different surface. These cuts remove a great deal of the waste and leave a rough, recognizable shape so you can begin carving the details almost immediately.

This same procedure will also generate finished sculptures if the shapes are abstract or require few details. Cut two different views in two different surfaces of the workpiece. Then, instead of removing additional stock with carving chisels, sand the sawed surfaces smooth. (*See Figures 6-5 through 6-7.*)

You can also create striking freehand sculptures using a variation of the compound cutting technique. For this procedure, you don't need to draw a pattern or even preplan the cuts. Just cut a contour — *any* contour — in the face of a rectangular block. Tape the waste back to the workpiece, turn the block on edge, and cut a second contour. Turn the pieces so the outside corners all face in and glue them together. (*See Figures 6-8 and 6-9.*) The resulting piece will look as if the board were bent in an intricate three-dimensional curve.

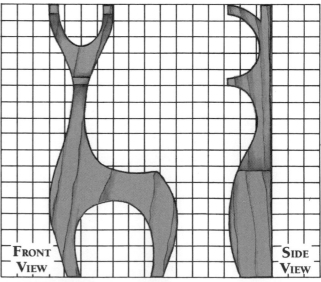

6-5 You can create simple sculptures with compound cuts. First, draw an outline of the subject from two different angles, 90 degrees apart. These patterns show a reindeer from the front and the side.

6-6 Trace the simplest of the two outlines on the workpiece — in this case the side view of the reindeer. Cut the outline and tape the waste back to the block. Turn the block 90 degrees, trace the second outline (the front view), and cut it.

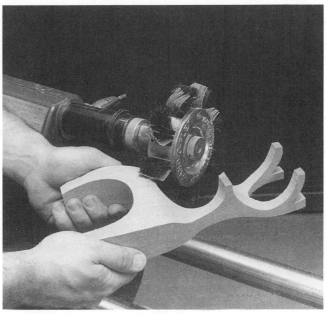

6-7 Remove the waste and discard the tape, then sand the sawed surfaces. If you want to preserve the hard edges, use a drum sander and a strip sander. If you want to soften the edges, making the piece appear more contoured, use a flap sander, as shown.

6-8 You can make freehand sculptures by cutting a series of curves. First, label the corners of a rectangular workpiece. Cut a contour down the length of the wood, sawing it from end to end. You don't have to plan the cut; any curved shape will do as long as you don't cut the wood into more than two pieces. Tape the pieces back together, turn the block 90 degrees, and cut a second contour from end to end. After removing the tape, you should have four pieces of wood. Turn each piece so its corner faces the others and glue the pieces together.

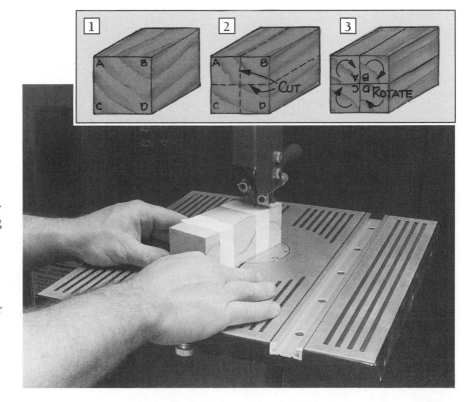

6-9 After the glue dries, sand the sawed surfaces. The resulting sculpture will look as if you bent the rectangular board — or partially melted it. These pieces make unique legs and posts for contemporary furniture.

MAKING CABRIOLE LEGS

The most common application for compound cutting is making *cabriole legs*. These are S-shaped legs for classic and traditional furniture. Cabriole legs come in many sizes and shapes, but all of them have several features in common — post, knee, ankle, and foot. (*See Figure 6-10.*)

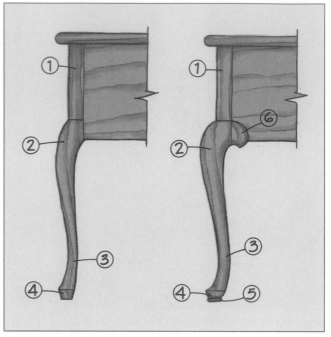

6-10 All cabriole legs have a straight section or *post* (1) at the top — this is where the leg is joined to the assembly. The shaped portion begins at the *knee* (2), which curves out. The curve reverses at the *ankle* (3), then ends in a *foot* (4). Some cabriole legs may also have a *pad* (5) on the bottom of the foot. Others have decorative ears or *transition blocks* (6) on the sides, near the tops of the knees.

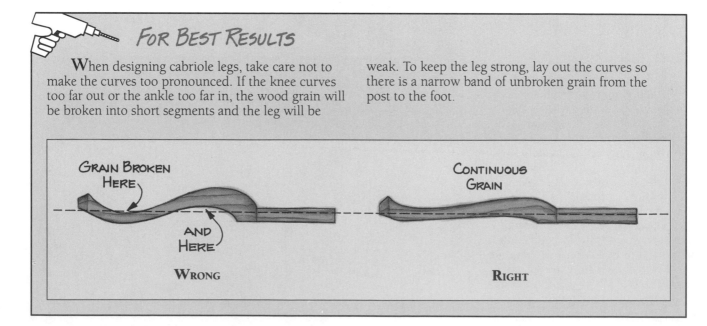

FOR BEST RESULTS

When designing cabriole legs, take care not to make the curves too pronounced. If the knee curves too far out or the ankle too far in, the wood grain will be broken into short segments and the leg will be weak. To keep the leg strong, lay out the curves so there is a narrow band of unbroken grain from the post to the foot.

GRAIN BROKEN HERE

AND HERE

CONTINUOUS GRAIN

WRONG

RIGHT

Some cabriole legs also have "ears" or *transition blocks* which make the curved legs seem to flow from the parts to which they're attached. The absence or presence of these transition blocks dictates which compound cutting method you must use to cut the legs. If the legs have no transition blocks, you can use the basic compound cutting technique. Trace the pattern for the cabriole leg on two adjacent sides of the leg stock. Cut one side, tape the waste back to the stock, and cut the second side. (*SEE FIGURES 6-11 AND 6-12.*)

6-11 **To cut a cabriole leg *without*** transition blocks, first make a leg pattern from posterboard or thin scraps of plywood or hardboard. Trace the pattern on two adjacent sides of the leg stock. Flip the pattern over when marking the second side so each layout is a *mirror image* of the other. **Note:** Be very careful that parts of the layouts — post, knee, ankle, and foot — line up with each other horizontally. The most critical part is the transition from the post to the knee. To make sure that it's aligned, mark a horizontal line on the pattern through this point. Carefully measure the position of the post/knee transition on both surfaces of the leg and, using a square, mark horizontal lines on the stock. When positioning the pattern on the stock, align the horizontal lines.

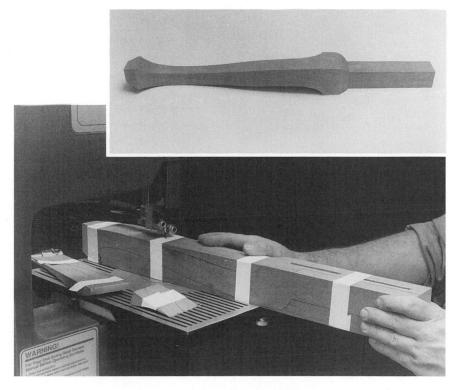

6-12 **Cut one side of the leg and** tape the waste back to the block. Turn the block and cut the second side. When you remove the waste, you'll have a simple cabriole leg.

If the legs do have transition blocks, the procedure is slightly more complex. You must glue the blocks to the leg stock, then trace the pattern on all the parts — legs *and* ears. When cutting, you must cut through both parts at times. (*SEE FIGURES 6-13 THROUGH 6-17.*) All the while, be sure to keep the upper blade guide the proper distance from the stock. This means that you occasionally will have to turn off the band saw, wait for the blade to coast to a stop, adjust the blade guide, then restart the saw.

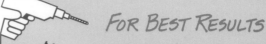

FOR BEST RESULTS

Always cut mortises and any other joinery in the leg stock *before* you cut the cabriole shape. It's very difficult to make accurate joinery cuts afterwards.

6-13 **To cut a cabriole leg *with*** transition blocks, make *two* patterns — one for the leg and one for the blocks. Glue the stock for the blocks to the leg stock, making sure that the grain direction matches. Trace the patterns on the *inside* surfaces (those surfaces that will face in on the assembled piece) of the legs and transition blocks. Once again, the patterns should be mirror images of each other. Note that the transition blocks are marked on two sides, just like the legs. One side shows the shape of the ear; the other, the shape of the knee.

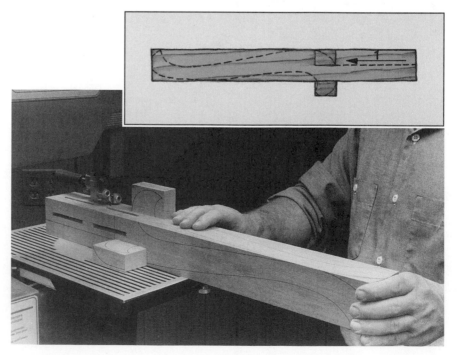

6-14 **Turn the leg so one inside** surface faces up. Cut the post until the upper blade guide bumps into the transition block. Turn off the band saw, let the blade coast to a stop, and raise the blade guide to clear the block. Finish cutting the post up to the transition block.

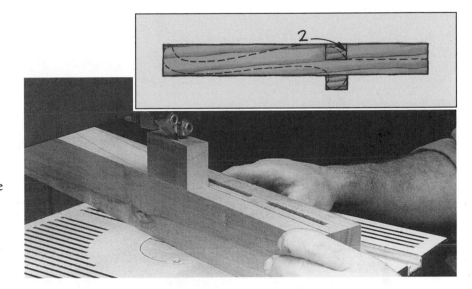

6-15   While the upper blade guide is raised above the transition block, cut the knee shape. This will free a large chunk of waste from the upper portion of the leg. Turn off the saw, let the blade stop, and lower the blade guide to its original position.

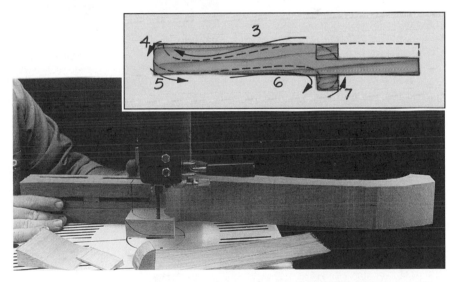

6-16   Cut the ankle and foot in the leg stock, then cut the ear shape in the other transition block. With most cabriole designs, you should be able to do this without raising the upper blade guide again. However, if the blade guide bumps into the transition block, do what you did before — turn the band saw off, let the blade stop, and raise the guide. Cut past the transition block, turn off the saw, and lower the guide. As you cut, remember to save the waste.

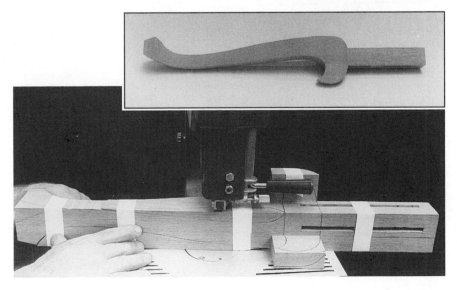

6-17   Tape the waste back to the leg and transition blocks. Turn the workpiece so the other inside surface faces up and repeat the same sequence of cuts, raising and lowering the upper blade guide as needed. When you completely remove the waste, you'll have a classic cabriole leg with ears.

Once you have cut the cabriole leg, you may want to round-over the corners of the knee, ankle, and foot. Round the knee and ankle by shaving the corners with a drawknife or spokeshave. (*SEE FIGURES 6-18 AND 6-19.*) Round the foot (and the pad, if you've made one) with a rasp and a file. (*SEE FIGURE 6-20.*)

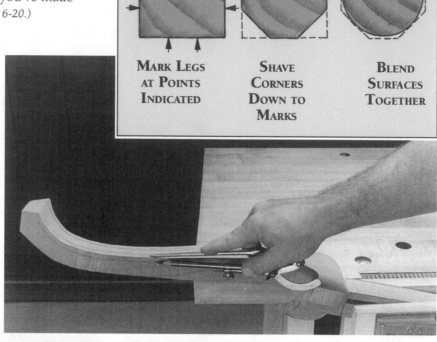

MARK LEGS AT POINTS INDICATED   SHAVE CORNERS DOWN TO MARKS   BLEND SURFACES TOGETHER

6-18 To round the knee and ankle of a cabriole leg accurately, first shave the square corners to give these portions an *octagonal* cross section, then blend the surfaces together to make the leg truly round. Begin by drawing guidelines on the rough-sawn leg. Using a compass or a marking gauge, scribe lines parallel to the corners, as shown. The distance between the guideline and the corners should be approximately one-third the width of the leg at the ankle (its narrowest point). For example, if the ankle is ³⁄₄ inch wide, draw the guidelines ¹⁄₄ inch from the corners.

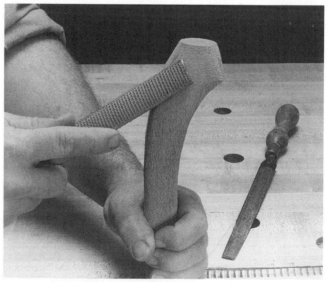

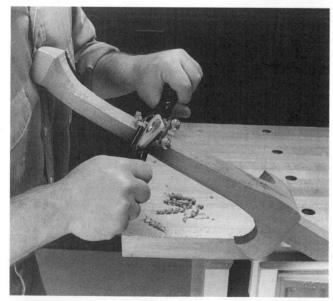

6-19 Using a drawknife or a spokeshave, shave each corner down to the guidelines to make the knee and ankle octagonal. Then go back over these areas with the same tool and blend the surfaces together, rounding them over in the process. Also, blend the ankle into the foot.

6-20 The procedure is similar for rounding the foot — draw guidelines parallel to the corners, cut the corners down to the guidelines to make the foot octagonal, then blend the surfaces to make it round. However, you must cut with rasps and files, since the surfaces aren't large enough to use spokeshaves or drawknives.

CUTTING CIRCLES

TURNING THE WORKPIECE ON A PIVOT

You can easily freehand a circle on the band saw, cutting the arc by simply following a circular layout line. However, you can cut circles more precisely by rotating the workpiece on a *pivot* as you feed it into the blade. This pivot need be nothing more than the point of a screw driven through a scrap of wood. *(SEE FIGURE 6-21.)* The distance from the blade to the pivot determines the radius of the circle.

When you use a pivot, you are making a guided cut. Because of this, the position of the pivot relative to the blade and the lead line is very important. Place the pivot even with the blade teeth along an imaginary line drawn perpendicular to the lead line. (For instructions on how to find and mark the lead line, see "Making Guided Cuts" on page 44.)

Mark the circle so one edge of the stock is tangent to the circumference — this will give you a place to start cutting. If the circle is very large, remove most of the waste from the stock with a circular saw or a saber saw. Otherwise, the waste may hit the band saw column as you rotate the workpiece on the pivot. *(SEE FIGURE 6-22.)*

Drill a small pivot hole in the underside of the stock, where it won't show on the assembled project. Place the workpiece so the hole slips over the pivot and the tangent edge is against the blade. Holding the workpiece in place, turn on the band saw. Rotate the workpiece on the pivot, cutting the circle.

TRY THIS TRICK

If you design a pivot so it can be attached to a band saw *and* a disk sander, you can use the same fixture to cut circles and sand the sawed edges.

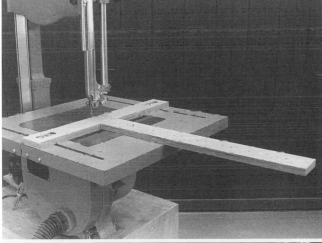

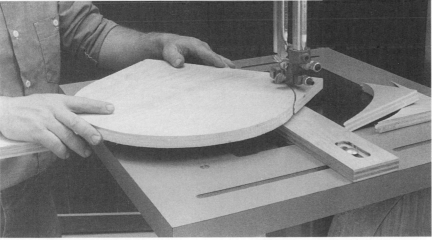

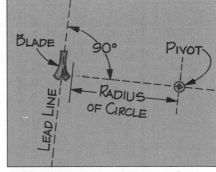

6-22 To position the pivot, first find the blade lead and mark the lead line. Using a wax pencil, draw another line perpendicular to the lead line out from the blade teeth. Place the pivot along this second line. Lay out the circle on the stock with one edge of the blade tangent to the circumference. Or, make a cut tangent to the circumference. When you start cutting, begin where the edge meets the circumference and rotate the workpiece on the pivot.

6-21 The pivot can simply be a screw in a scrap of wood that is clamped to the band saw table. If you want to make a fixture that's more versatile and easier to use, this circle-cutting jig offers an *adjustable* pivot that you can position almost any distance (up to 24 inches) from the blade to cut both large and small circles. For plans and instructions on how to make this accessory, see "Circle-Cutting Jig" on page 84.

BAND SAW JOINERY

MAKING FINGER JOINTS

You can use the band saw to cut a variety of joints, in particular *finger joints* and *dovetail joints*. The basic procedure for making both of these interlocking joints is very similar:

1. Carefully lay out the fingers, tails, or pins on the boards to be joined.
2. Cut the cheeks (sides) of the fingers, tails, or pins.
3. Clean out the waste between them.

When making finger joints, clamp the boards to be joined face to face and lay out the joint across the ends of both boards. (*SEE FIGURE 6-23.*) Remove the clamps and mark the fingers on the faces of the boards. (*SEE FIGURE 6-24.*) Cut the cheeks of the fingers, using a stop to halt the cut at the proper depth. (*SEE FIGURE 6-25.*) Finally, remove the waste and "nibble" the notches between the fingers to the proper depth. (*SEE FIGURES 6-26 AND 6-27.*)

CIRCLE-CUTTING JIG

This simple T-shaped fixture is just two pieces of wood joined with a lap joint. The pivot arm holds the pivot a specific distance from the blade, while the crossbar can be clamped to the band saw table.

If you cut two slots in the crossbar, as shown, you can attach the fixture to the small auxiliary table shown on page 47.

To use the jig, first decide the radius of the circle

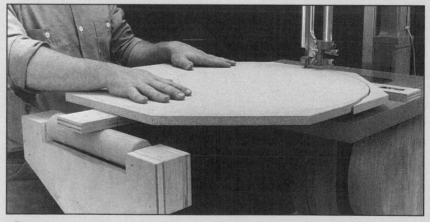

1 **Mount the circle-cutting jig** on the band saw table with the pivot arm pointing away from the column. Carefully align the arm square to the lead line and make sure the pivot is even with the blade teeth. Move the jig closer to or farther away from the blade until the distance between the blade teeth and the pivot is equal to the radius of the circle you wish to cut. Then clamp the crossbar so the jig won't move.

2 **Prepare a workpiece, mount** it on the pivot, and rotate it slowly as you cut it. If the workpiece is especially large or heavy, use a saw stand to help support the protruding end of the arm.

6-23 To lay out a finger joint, clamp the boards together face to face with the edges and adjoining ends flush. Using a square, mark the fingers across both ends at the same time. With a pencil, shade in the waste on each board. Note that the fingers don't have to be all the same size. This is a major advantage of making finger joints with a band saw — you can vary the size of the fingers for decorative effect. In this particular finger joint, the fingers will become progressively larger across the width of the boards.

you wish to cut. Then drive a flathead screw up through one of the holes in the pivot arm. The distance between the screw and the outside edge of the crossbar should be *slightly* less than the radius

of the circle. The point of the screw should protrude $\frac{1}{4}$ to $\frac{1}{2}$ inch from the leg — this will serve as the pivot.

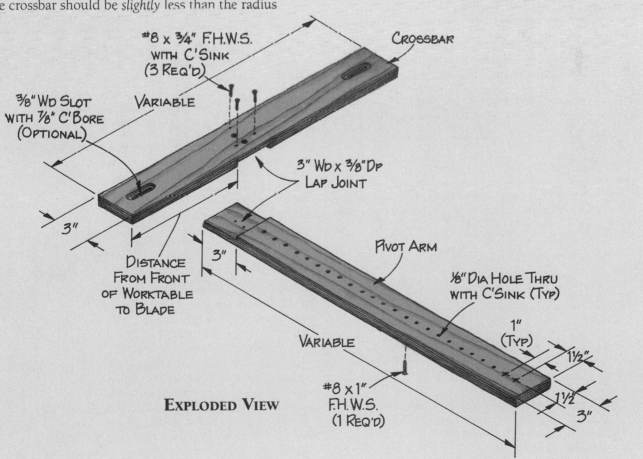

EXPLODED VIEW

#8 × ¾" F.H.W.S. WITH C'SINK (3 REQ'D)

CROSSBAR

⅜" WD SLOT WITH ⅞" C'BORE (OPTIONAL)

VARIABLE

3"

3" WD × ⅜" DP LAP JOINT

DISTANCE FROM FRONT OF WORKTABLE TO BLADE

3"

PIVOT ARM

⅛" DIA HOLE THRU WITH C'SINK (TYP)

VARIABLE

1" (TYP)

1½"

#8 × 1" F.H.W.S. (1 REQ'D)

1½"

3"

6-24   Remove the clamps and take the boards apart. Scribe the baselines of the fingers with a marking gauge, then mark the cheeks with a square. Once again, shade the waste between the fingers with a pencil. This is very important — when you start cutting, it's easy to lose track of what's waste and what's a finger.

6-25   Clamp a scrap to the band saw table *behind* the blade to serve as a stop. The distance between this stop and the blade teeth should be equal to the length of the fingers you plan to make. Cut the cheek of each finger, shaving the *waste* sides of the layout lines.

6-26   To remove the waste, make multiple cuts to divide the waste into narrow slivers. Then go back and make additional cuts to remove the slivers. This will create the notches between the fingers. (*Blade guide raised for clarity.*)

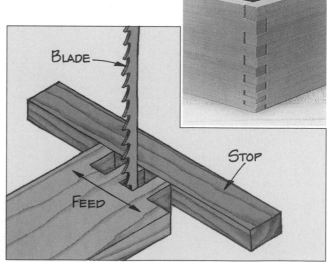

6-27   Using the stop as a guide, feed the workpiece sideways to nibble away the last remaining bits of waste at the bottoms of the notches. This will also square the bottoms of the notches to the cheeks and cut all the notches to precisely the same depth. Finally, fit the adjoining boards together. If the finger joint seems tight, remove a little stock from the cheeks of the fingers with a flat file.

MAKING DOVETAIL JOINTS

To make a dovetail joint, lay out the pins on one board. Clamp the adjoining boards together face to face and transfer the positions of the pins to the end of the other board. (SEE FIGURE 6-31.) Remove the clamps and mark the tails on the second board. (SEE FIGURE 6-29.) Cut the cheeks of the pins and remove the waste, using a special jig fastened to the band saw table. (SEE FIGURE 6-30.) Remove the jig and cut the cheeks of the tails. Clean out the waste between the tails and fit the tails and pins together. (SEE FIGURE 6-31.)

FOR BEST RESULTS

Use an awl or a marking knife to lay out finger joints and dovetail joints. These tools leave a much finer line than a pencil, and this makes it easier to be accurate. When cutting the joints, use a wide blade (at least 1/2 inch wide) with standard or skip teeth for the smoothest, straightest possible cut.

END OF PIN BOARD — NARROW DIMENSION OF PINS — END OF TAIL BOARD

6-28 Designate which of the two adjoining boards will have the dovetails and which will have the pins. Using a marking gauge, scribe baselines for the tails and pins on both boards. Then lay out the pins on the pin board, using a square and a sliding T-bevel. Clamp the two boards face to face with the edges and the adjoining ends flush. The *narrow* dimension of the pins should butt against the tail board, as shown in the drawing. Using a square, transfer this dimension from the pin board to the tail board.

6-29 Lay out the tails to match the pins, using the sliding T-bevel. (It should still be set to the same angle that you used to lay out the pins.) Shade the waste between the pins and tails with a pencil. **Note:** Dovetails should slope at an angle between 7 and 14 degrees, and most are made between 8 and 10 degrees. The tails and pins shown slope at 10 degrees.

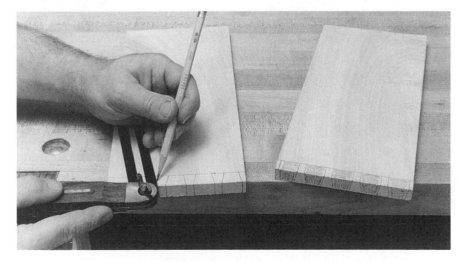

6-30   Cut the cheeks of the pins.
To do this, you must tilt the table to the right *and* left. If your band saw table won't tilt as far as necessary in both directions, make the tilting worktable shown. (For plans and instructions, refer to "Tilting Work-table for Cutting Dovetails" below.) Attach this to your band saw table and make sure the degree of table tilt matches the slope of the dovetails. Adjust the distance between the blade teeth and the stop at the back of the table so it's equal to the length of the pins. Cut the right-sloping cheeks of the pins with the table tilted right, and the left-sloping cheeks with the table tilted left. Slice and nibble the waste from between the pins, using the stop as a guide. **Note:** Always cut to the waste side of the layout lines.

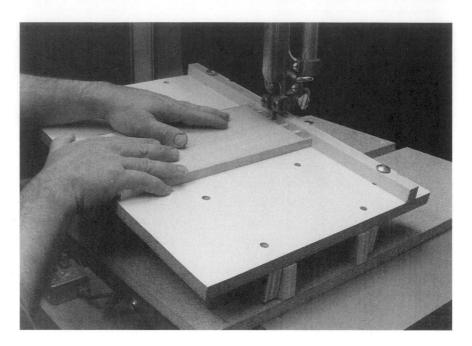

TILTING WORKTABLE FOR CUTTING DOVETAILS

When making the pins in a dovetail joint, you must tilt your table to the right to cut half of the pin cheeks, then back to the left to cut the other half. The trouble is, not all band saw tables tilt to the left. And those that do may not tilt as far as necessary. To solve this problem, make a tilting worktable that clamps to your band saw table.

The tilting worktable works like a teeter-totter. The work surface is attached at the front and the back to pivots, and these pivots are attached to a base. The back (outfeed) edge of the work surface is notched to fit around the blade — this helps support the stock as it's cut. A detachable stop controls the length of each cut.

The tilt of the worktable must match the slope of the dovetails you want to cut. As designed, the fixture will tilt 10 degrees to the left and right. If you want to cut dovetails at some other angle, you must adjust the design.

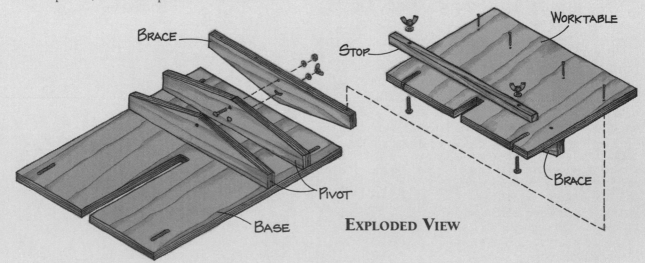

EXPLODED VIEW

6-31 Detach the tilting worktable from the band saw and cut the cheeks of the tails. Remove the waste from between the tails by slicing and nibbling. Because these cuts are angled, you won't be able to use a stop to guide them, so be careful to nibble up to the baseline and no farther. If you wish, clean up the bases of the notches with a triangular file. *(Blade guide raised for clarity.)*

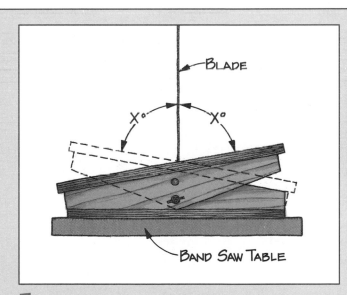

1 **When you mount the tilting** worktable on your band saw, check that the tilt is *symmetrical*. Measure the angle between the blade and the work surface with the table tilted left and right — the two angles should be equal. If the tilt is asymmetrical, equalize it by tilting the band saw table.

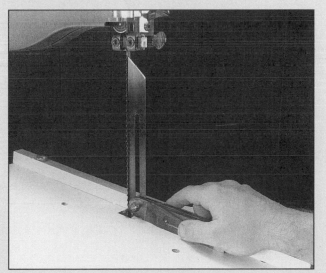

2 **To make sure that the slope** of the dovetails and the tilt of the worktable are the same, use the worktable to set the sliding T-bevel before you lay out the tails and pins. Tilt the table to the right, place the base of the T-bevel on the work surface, and adjust the arm to lay flat against the blade body. Double-check the angle by tilting the table to the left and moving the T-bevel to rest against the other side of the blade.

(continued) ▷

TILTING WORKTABLE FOR CUTTING DOVETAILS — CONTINUED

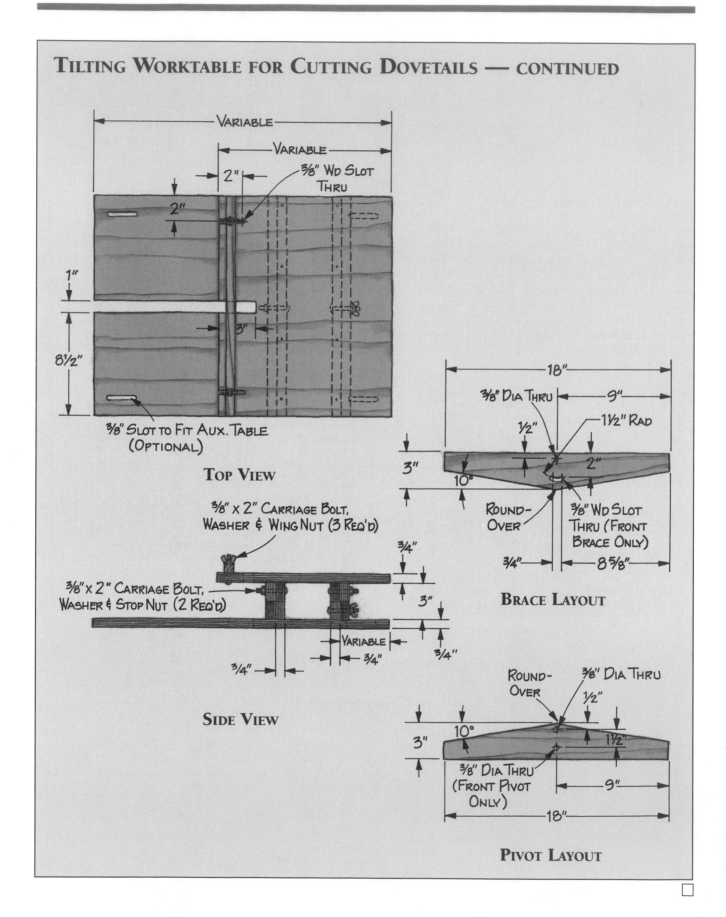

VARIABLE

VARIABLE

2"

3/8" WD SLOT THRU

2"

1"

8½"

3"

3/8" SLOT TO FIT AUX. TABLE (OPTIONAL)

TOP VIEW

3/8" x 2" CARRIAGE BOLT, WASHER & WING NUT (3 REQ'D)

3/8" x 2" CARRIAGE BOLT, WASHER & STOP NUT (2 REQ'D)

3/4"

3"

VARIABLE

3/4"

3/4"

3/4"

SIDE VIEW

18"

9"

3/8" DIA THRU

1½" RAD

½"

3"

2"

10°

ROUND-OVER

3/8" WD SLOT THRU (FRONT BRACE ONLY)

3/4"

8 5/8"

BRACE LAYOUT

ROUND-OVER

3/8" DIA THRU

½"

3"

10°

1½"

3/8" DIA THRU (FRONT PIVOT ONLY)

9"

18"

PIVOT LAYOUT

PROJECTS

7

NOAH'S BAND-SAWED ARK

During the nineteenth century, parents gave their children special toys on Sunday to keep them calm and peaceful. One of the most popular "Sunday toys" was Noah's Ark — a set of animals and a boat-shaped box to keep them in. This not only inspired quiet play but had a biblical theme. Today, Noah's Arks are both great toys and highly valued works of folk art.

This particular Noah's Ark is a band saw project from start to finish. *All* the shapes you see — animals, the hull, the cabins, even the rounded roof — were cut on the band saw. The only other power tools used were sanders.

This ingenious band saw project was designed by folk artist Mary Jane Favorite. As in all of Ms. Favorite's arks, she has thoughtfully provided separate quarters for the skunks — a "skunk's nest" perched atop the roof. The other animals occupy three compartments inside the ark. The roof lifts off of the upper story of the cabin to reveal the first compartment, then the upper cabin lifts off the lower cabin to access the second. Finally, the lower cabin lifts off to show the third compartment in the hull.

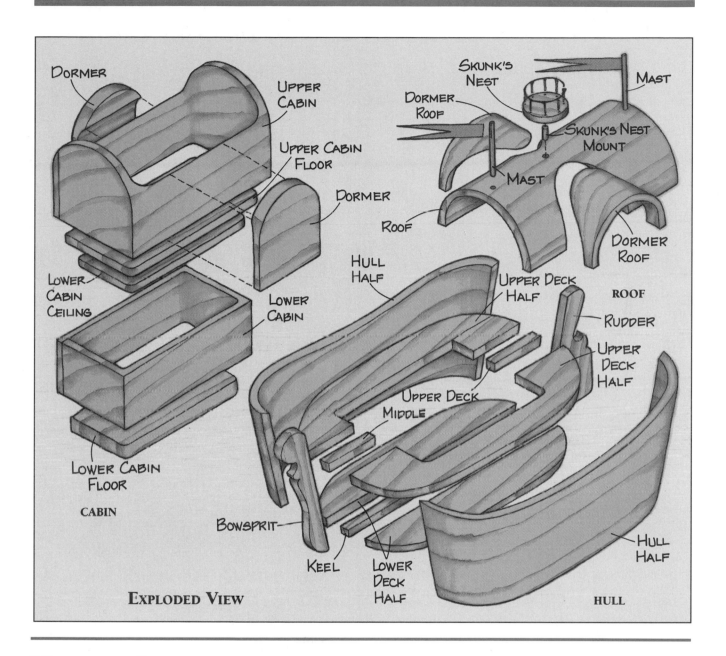

EXPLODED VIEW

Labels in figure:
DORMER
UPPER CABIN
UPPER CABIN FLOOR
DORMER
LOWER CABIN CEILING
LOWER CABIN
LOWER CABIN FLOOR
CABIN
SKUNK'S NEST
DORMER ROOF
MAST
SKUNK'S NEST MOUNT
MAST
ROOF
DORMER ROOF
ROOF
RUDDER
UPPER DECK HALF
UPPER DECK HALF
HULL HALF
UPPER DECK
MIDDLE
BOWSPRIT
KEEL
LOWER DECK HALF
HULL HALF
HULL

MATERIALS LIST (FINISHED DIMENSIONS)

Parts

If you have a band saw with at least a 10″ depth of cut:

A. Cabin block   6″ x 9″ x 10″
B. Hull block   8½″ x 8½″ x 18″

If you have a band saw with at least a 6″ depth of cut:

A. Cabin block   3⅝″ x 5⅜″ x 6″
B. Hull block   5⅛″ x 5⅛″ x 10⅞″

In addition, you'll need:

C.   Masts (2)   ¼″ dia. x 2″
D.   Skunk's nest mount   ¼″ dia. x ½″

Note: The animals, animal bases, and skunk's nest are all cut from scraps from the cabin block and hull block.

Hardware

1″ Wire brads (12)
18-gauge wire (6½″)
Aluminum flashing or stiff canvas (two ¾″ x 5″ pieces)

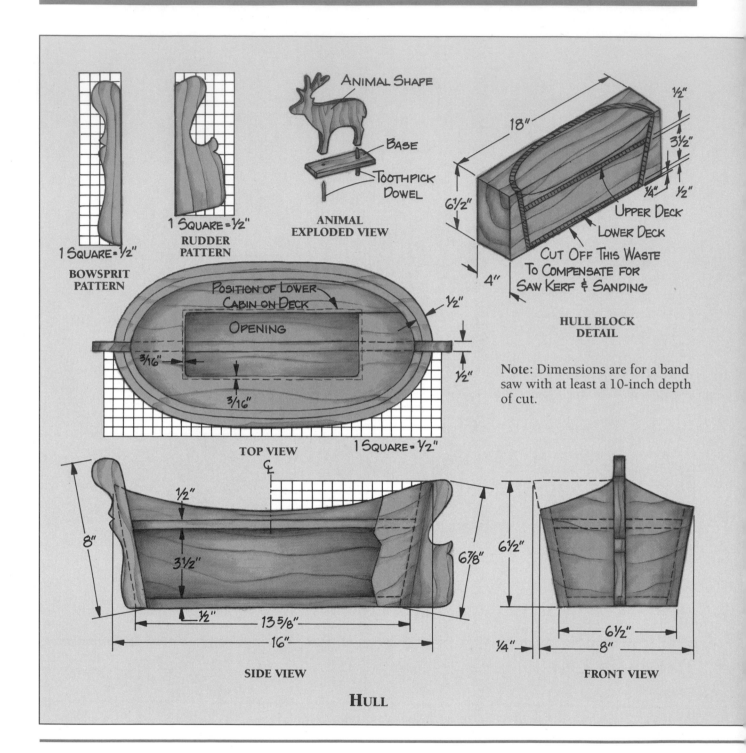

BOWSPRIT
PATTERN

1 SQUARE = ½"

RUDDER
PATTERN

1 SQUARE = ½"

ANIMAL SHAPE

BASE

TOOTHPICK
DOWEL

ANIMAL
EXPLODED VIEW

18"

6½"

4"

½"

3½"

¼" ½"

UPPER DECK

LOWER DECK

CUT OFF THIS WASTE
TO COMPENSATE FOR
SAW KERF & SANDING

HULL BLOCK
DETAIL

POSITION OF LOWER
CABIN ON DECK

OPENING

½"

½"

3/16"

3/16"

TOP VIEW

1 SQUARE = ½"

Note: Dimensions are for a band
saw with at least a 10-inch depth
of cut.

8"

½"

3½"

½"

13 5/8"

16"

6 7/8"

SIDE VIEW

6½"

¼"

6½"

8"

FRONT VIEW

HULL

PLAN OF PROCEDURE

1 Prepare the stock. The parts of the cabin and the hull are cut from two large blocks of wood, then glued back together. The grain direction of the blocks must be parallel so all the parts of the assembled ark will expand and contract evenly, in the same direction.

The size of the ark that you make depends on the maximum depth of cut of your band saw. The dimensions shown in the working drawings will only work if your saw will cut stock *at least* 10 inches thick. If, like most medium-size band saws, your machine has

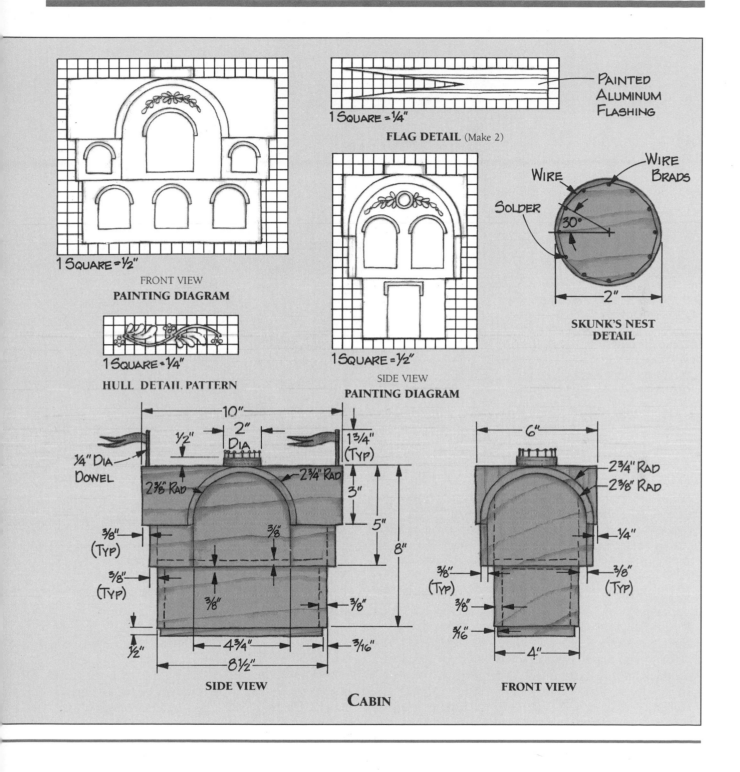

1 SQUARE = ½"

FRONT VIEW
PAINTING DIAGRAM

1 SQUARE = ¼"

HULL DETAIL PATTERN

1 SQUARE = ¼"

FLAG DETAIL (Make 2)

PAINTED
ALUMINUM
FLASHING

1 SQUARE = ½"

SIDE VIEW
PAINTING DIAGRAM

WIRE
WIRE
BRADS
SOLDER
30°
2"

**SKUNK'S NEST
DETAIL**

CABIN

SIDE VIEW

10"
½"
2"
DIA
1¾"
(TYP)
¼" DIA
DOWEL
2⅜" RAD
2¾" RAD
3"
5"
8"
⅜"
(TYP)
⅜"
⅜"
(TYP)
⅜"
⅜"
4¾"
½"
3/16"
8½"

FRONT VIEW

6"
2¾" RAD
2⅜" RAD
¼"
⅜"
(TYP)
⅜"
(TYP)
⅜"
3/16"
4"

a 6-inch depth of cut, reduce all the measurements to 60 percent. For example, if the plans show 5 inches, change this to 3 inches.

Glue up the blocks from a soft, light-colored wood such as basswood, white pine, alder, or juniper. You need a soft wood so you can use a fairly narrow blade to make very thick cuts. Otherwise, you won't be able to cut the curves needed. And a light-colored wood is easier to paint.

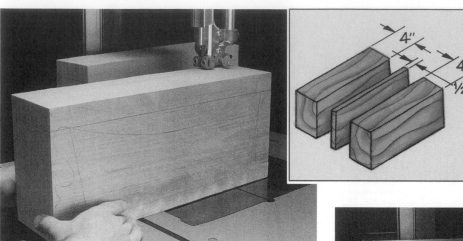

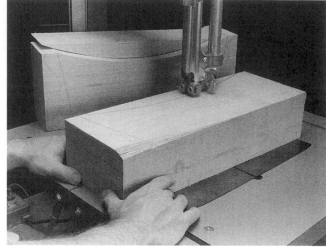

7-1 Split the hull block into three pieces. The *middle* piece should be ½ inch wide. This piece will later become the bowsprit, rudder, keel, and upper deck middle. The two *outside* pieces should each be about 4 inches wide. These will become the hull halves, lower deck halves, and upper deck halves.

2 Cut the parts of the hull. The parts of the hull must be cut in the proper order. It's *extremely* important that you follow this sequence *precisely*. See the *Exploded View* if you're not sure what part the instructions are referring to. Make the cuts in this order:

- Split the hull block in three pieces. (SEE FIGURE 7-1.)

- Resaw the two *outside* hull pieces so they are 6½ inches tall.

- Cut the gunwale curves in the two outside hull pieces. (SEE FIGURE 7-2.)

- Cut the hull halves from the two outside pieces with the band saw table tilted 10 degrees. (SEE FIGURE 7-3.)

- Cut the shapes of the bowsprit and rudder from the middle hull piece. (SEE FIGURE 7-4.)

- Resaw the scraps from all three pieces to make the lower deck halves, upper deck halves, keel, and upper deck middle. (SEE FIGURE 7-5.)

Note: All of the hull parts on this ark were cut with a ³/₈-inch standard-tooth blade. However, because you can't cut tight curves with this blade, you must do quite a bit of nibbling to make the shapes of the rudder and bowsprit. You may wish to change to a ¹/₈-inch standard-tooth blade to cut these two parts.

7-2 Resaw the 4-inch-wide hull pieces so they are just 6½ inches tall. Lay out the curve of the hull on top of these pieces, as shown in the *Hull/Top View*. Then lay out the curve of the gunwale (top edge of the hull), as shown in the *Hull/Side View*. Cut the gunwale curve in both pieces, saving the waste.

3 Assemble the hull. Using a drum sander and a disk sander, sand all the parts to remove the saw marks. Then glue the hull halves, rudder, and bowsprit together.

When the glue dries on the hull assembly, lay the lower deck parts in place and mark a line across all three parts. Glue the lower deck halves and keel together, using the mark you've made to properly align the parts. Repeat this procedure to assemble the upper deck parts.

Let the glue dry on the deck assemblies and glue the lower deck in the hull. However, do *not* glue the upper deck in place yet.

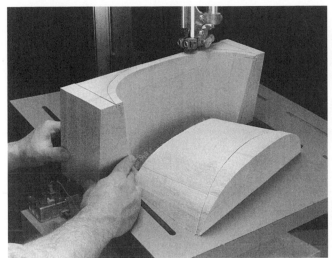

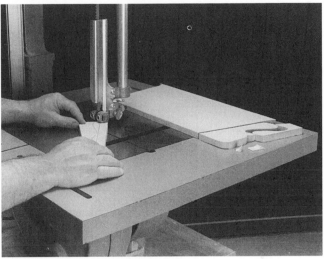

7-3 Reattach the waste to the 4-inch-wide blocks with double-faced carpet tape. Tilt the band saw table 10 degrees. Saw the *inside* curve of the hull halves, then the *outside* curve. As you cut, keep the inside of the hull facing *away* from the band saw throat. This way, the hull halves will slope in from the gunwale. Discard the scrap from the outside and top edges of the hull halves, but save the scrap from the inside.

7-4 Lay out the bowsprit and rudder on the ½-inch-wide *middle* piece, as shown in the *Bowsprit* and *Rudder* patterns. Be sure to mark these shapes on the wood precisely where the hull halves would join them if you were to leave the middle piece whole. Cut the shapes, saving the *inside* scrap.

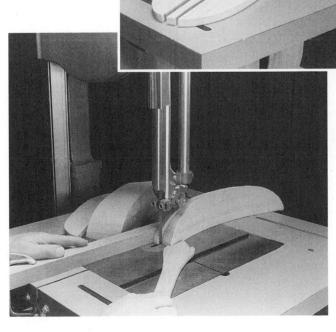

7-5 Resaw the three scrap pieces from inside the hull to make the lower deck halves, upper deck halves, keel, and upper deck middle. When you assemble these parts, they will make two oval-shaped decks, as shown in the inset photo. Because the outside curve of the decks matches the inside curve of the hull halves, they should fit snugly when you assemble the ark hull. However, because of the stock that's lost to sanding and the saw kerf, the decks will sit a little lower in the hull than they were in the uncut block. To compensate, cut ¼ inch of stock from the bottom surfaces of the three hull pieces, as shown in the *Hull Block Detail*. Then lay out and cut the deck parts.

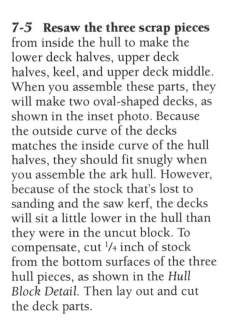

4 Cut the parts of the cabin.   As with the hull, the parts of the cabin must be cut in a specific sequence:

■ Split the cabin block in two pieces, from which you'll cut the upper and lower cabin parts. (*SEE FIGURE 7-6.*)

■ Cut the *inside* curve of the roof in the upper cabin piece. (*SEE FIGURE 7-7.*)

■ Cut the dormers from the roof piece. (*SEE FIGURE 7-8.*)

■ Resaw the roof piece to size, then cut the *outside* curve of the roof. (*SEE FIGURE 7-9.*)

■ Cut the dormer roofs from the roof scrap. (*SEE FIGURE 7-10.*)

■ Resaw the lower cabin piece to size, then cut out the interior of the lower cabin. (*SEE FIGURE 7-11.*)

■ Resaw the upper cabin piece and dormers to size, then glue the dormers on the upper cabin. (*SEE FIGURE 7-12.*)

■ Cut out the interior of the upper cabin. (*SEE FIGURE 7-13.*)

■ Resaw the scraps from the interior of the upper and lower cabin to make the upper cabin floor and lower cabin ceiling. (*SEE FIGURE 7-14.*)

■ Cut the lower cabin floor from the upper deck assembly. (*SEE FIGURE 7-15.*)

Note: All the surfaces on the cabin of this ark were cut with a $3/8$-inch standard-tooth blade *except* for the interiors of the upper and lower cabins and the opening in the upper deck. These were cut with a $3/16$-inch standard-tooth blade to make the tight turns at the corners.

7-7 Lay out the roof, dormers, and dormer roofs on the upper cabin piece. Also draw several lines across the bottom of this piece — later, these will help you position the dormers on the upper cabin. Stand the piece on end, then saw the sides of the upper cabin and the *inside* curve of the roof in one long cut. This will separate the roof piece from the upper cabin.

7-6 Resaw the cabin block into two pieces, one 3 inches tall and the other about $5\frac{1}{2}$ inches tall. You'll make the lower cabin and lower cabin ceiling from the smaller piece, and all the other cabin parts from the larger one.

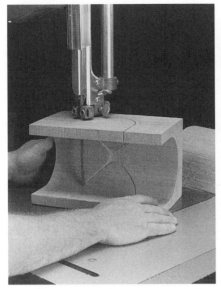

7-8 Rest the roof piece on its side and cut the dormers. *Be extremely careful* when doing this, because there's a lot of blade exposed inside the curve of the roof. Keep your fingers outside that curve at all times.

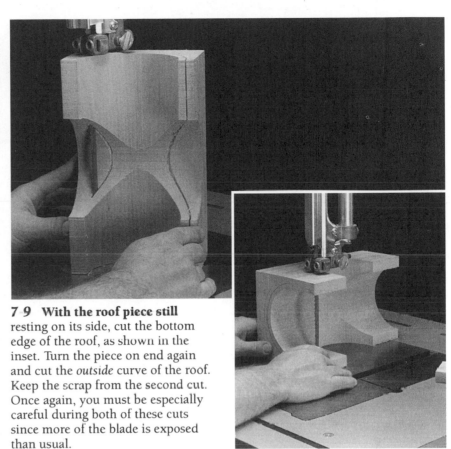

7-9 With the roof piece still resting on its side, cut the bottom edge of the roof, as shown in the inset. Turn the piece on end again and cut the *outside* curve of the roof. Keep the scrap from the second cut. Once again, you must be especially careful during both of these cuts since more of the blade is exposed than usual.

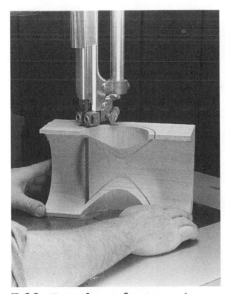

7-10 Turn the roof scrap on its side and cut the dormer roofs in the same manner that you cut the dormers. Again, keep your hands outside the curve of the stock — there's a lot of blade exposed during this procedure.

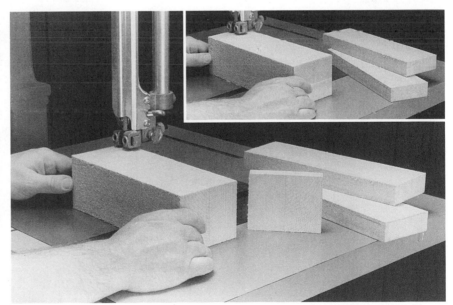

7-11 Lay out the lower cabin walls on the top surface of the lower cabin piece. The lower cabin should be 8½ inches long and 4 inches wide, and each wall should be ⅜ inch thick. Resaw the sides and ends of the lower cabin piece, cutting it to size as shown in the inset. Then, starting at one end of the lower cabin, saw into the block and cut out the interior. Save the scrap from the interior.

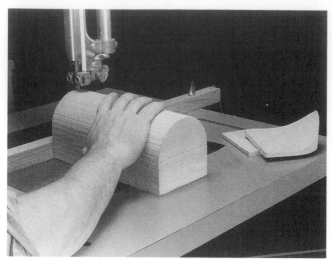

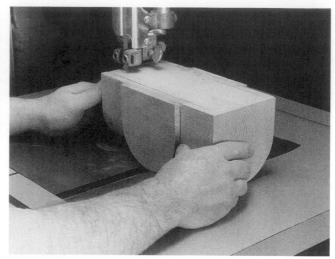

7-12   Resaw ¼ inch of stock from
the flat sides (the outside surfaces)
of the dormers, and ⅜ inch from
each end of the upper cabin piece.
Sand the saw marks from the upper
cabin and the dormers, then glue the
dormers to the sides of the upper
cabin. The position of these dormers
is critical, so use the marks you
made on the bottom of the upper
cabin piece to help align them.

7-13   Turn the upper cabin
assembly over so it's resting on its
top. Lay out the walls on the bottom
surface. The end walls are ⅜ inch
thick, and the side walls are ⅜ inch
thick near the ends and ¾ inch thick
where the dormers are attached.
Starting at one end, saw into the
block and cut out the interior of the
upper cabin. Save the scrap.

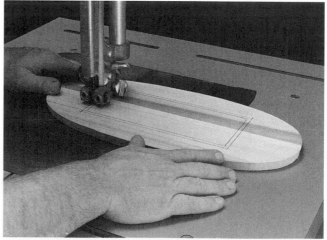

7-14   Resaw the scraps from the
interior of the upper and lower cabins.
Cut a ⅜-inch-thick slab from the bottom surface of the upper cabin scrap
to make the upper cabin floor, and
another ⅜-inch-thick slab from the
top surface of the lower cabin scrap
to make the lower cabin ceiling.

7-15   Center the lower cabin on
the upper deck assembly and trace
around the outside. Then lay out the
opening in the deck, as shown in the
Hull/Top View. This opening should
be ⅜ inch narrower and shorter than
the length and width of the lower
cabin. Saw into the deck assembly
from one end and cut out the opening. The scrap from the opening will
become the lower cabin floor.

5 **Glue the upper deck to the hull.** Glue the upper deck back together where you cut into the stock to make the opening. Let the glue dry, then sand the sawed edges of the opening. Glue the upper deck in place in the hull.

6 **Assemble the cabin.** Sand the sawed surfaces to remove the saw marks. Then glue the parts together in this order:

■ Glue the upper and lower cabin walls together where you sawed into the stock to cut out the interior.

■ Center the lower cabin floor on the lower cabin and glue the parts together.

■ Glue the upper cabin floor in the upper cabin. Let the glue dry, then center the lower cabin ceiling on the upper cabin floor and glue the parts together.

■ Glue the dormer roofs to the roof.

When the cabin and hull are completely glued together, you should have four subassemblies — hull, lower cabin, upper cabin, and roof. These must all fit together. The lower cabin fits in the opening in the hull; the upper cabin ceiling fits over the lower one; and the roof fits the contours of the upper cabin.

7 **Make the skunk's nest.** Resaw a 1/2-inch-thick slab from one of the many scraps left over from making the hull or cabin. From this, cut a 2-inch-diameter skunk's nest. Also cut a 1/2-inch length of 1/4-inch-diameter dowel to mount the nest on the roof.

Drill a 1/4-inch-diameter, 1/4-inch-deep hole in the center of the skunk's nest and a matching hole in the center of the roof, between the two dormer roofs. Drive wire brads partway into the skunk's nest, all around the perimeter. Solder a length of wire to the protruding ends of these brads, as shown in the *Skunk's Nest Detail.*

Glue the mounting dowel to the skunk's nest, but *don't* glue the nest to the roof yet. Wait until after you have painted both assemblies.

8 **Make the flags.** Cut 2-inch lengths of 1/4-inch-diameter dowel to make the masts for the flags. Using a 1/8-inch or 3/16-inch blade on the band saw, cut 1-inch-long slots in the top ends of the masts. Cut 3/4-inch-wide, 5-inch-long strips of aluminum flashing to make the flags, as shown in the *Flag Detail.* Prime the surface of the aluminum with zinc chromate primer — this will allow you to paint the metal with either water-base or oil-base paints. Let the primer dry, then fasten the flags in the mast slots with epoxy cement.

A SAFETY REMINDER

If this ark will be played with by small children, use stiff canvas instead of metal flashing to make the flags.

Drill 1/4-inch-diameter, 1/4-inch-deep holes near both ends of the roof to mount the flags. However, *don't* glue the flags in place until after you've painted both assemblies.

9 **Make the animals.** Cut the animals' shapes, as well as Mr. and Mrs. Noah, from the hull and cabin scraps. Resaw these scraps to make 1/4-inch-thick slabs. Lay out the shapes of the animals on the slabs and cut them with a 1/8-inch standard-tooth blade.

TRY THIS TRICK

To save time, *pad saw* the male and female animal shapes whenever you can. If the male and female profiles are the same, stick two pieces of wood together with double-faced carpet tape. Saw both pieces at the same time, then take them apart and discard the tape.

Also cut bases for the animals. The size of each base will vary with the size of the animal, but generally they should be about as long as the animal and wide enough to prevent the animal from tipping over. The taller the animal, the wider the base should be. Small animals, such as the skunks, only require 1/2-inch-wide bases. Taller animals, such as giraffes and elephants, need 1-inch-wide bases. Most animals will be perfectly steady on 3/4-inch-wide bases.

Sand the sawed edges of the animals and bases, then glue each animal to its base. If you wish, you can reinforce the glue joints with dowels made from toothpicks. Drill 1/16-inch-diameter holes through the bases and into the legs of the animals. Roll the ends of round toothpicks in glue and insert them in the holes. After the glue dries, use a sharp chisel to trim the toothpicks flush with the bases.

Note: There is only room for a limited number of animal patterns in this book, but you'll have space for lots and lots of animals in the three compartments in the ark. Should you want to make more animals than are shown here, consult field guides and other nature books for ideas.

10 Paint the ark and the animals.   Paint the ark, skunk's nest, flags, and animals to suit your fancy. If you want some ideas as to what colors to paint the ark, consult books on folk art and "country" decor. For references on how to paint the animals, look them up in field guides.

The ark and the animals shown are painted with artists' acrylic paints. These paints cover well, dry quickly, and are easy to clean up. The paints are low in toxicity, a plus if you're making this ark for children. They are also available in a wide range of colors, and you can easily mix them to create still more hues. If you need more working time, and toxicity isn't a concern, use artist's oils instead. These offer many of the advantages of acrylics, but they are slower to dry (and slightly more expensive).

After the paint has dried, glue the flags and the skunk's nest in place on the roof.

ANIMAL PATTERNS

RHINOCEROS

TIGER

FOX

LION

OSTRICH

LIONESS

BEAR

SWAN

GIRAFFE

ELEPHANT

1 SQUARE = ¼"

ANIMAL PATTERNS

8

QUEEN ANNE SIDE TABLE

Small tables such as these once were employed as sewing tables or work tables all around the home. Today, they're used for more leisurely pursuits. You find them beside sofas, easy chairs, and beds, holding lamps, books, and television remotes.

This small table displays the graceful cyma (S-shaped) curves that mark classic Queen Anne furniture. The sculpted contours of the cabriole legs blend with the curves at the bottoms of the aprons. All of these shapes, both three dimensional and two dimensional, were cut on a band saw.

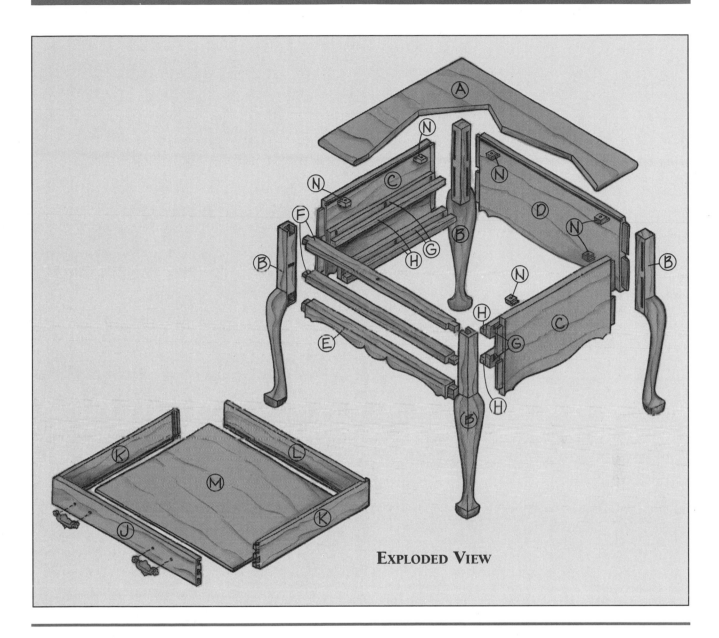

EXPLODED VIEW

MATERIALS LIST (FINISHED DIMENSIONS)

Parts

A.	Top	3/4" x 24" x 30"
B.	Legs (4)	2 1/2" x 2 1/2" x 26 1/4"
C.	Side aprons (2)	3/4" x 11" x 20"
D.	Back apron	3/4" x 11" x 26"
E.	Bottom rail	1 1/2" x 2" x 26"
F.	Top/middle rails (2)	3/4" x 1 1/2" x 26"
G.	Drawer guides (4)	3/4" x 1 1/2" x 18"
H.	Drawer supports (4)	3/4" x 3/4" x 18 3/4"
J.	Drawer fronts (2)	3/4" x 3 11/16" x 23 15/16"
K.	Drawer sides (4)	1/2" x 3 11/16" x 20 1/8"
L.	Drawer backs (2)	1/2" x 3 11/16" x 23 7/16"
M.	Drawer bottoms* (2)	1/4" x 19 1/4" x 23 7/16"
N.	Cleats (6)	1/2" x 1" x 1"

Make these parts from plywood.

Hardware

#8 x 1" Flathead wood screws (6)

#8 x 1 1/4" Flathead wood screws (13)

Drawer pulls (4)

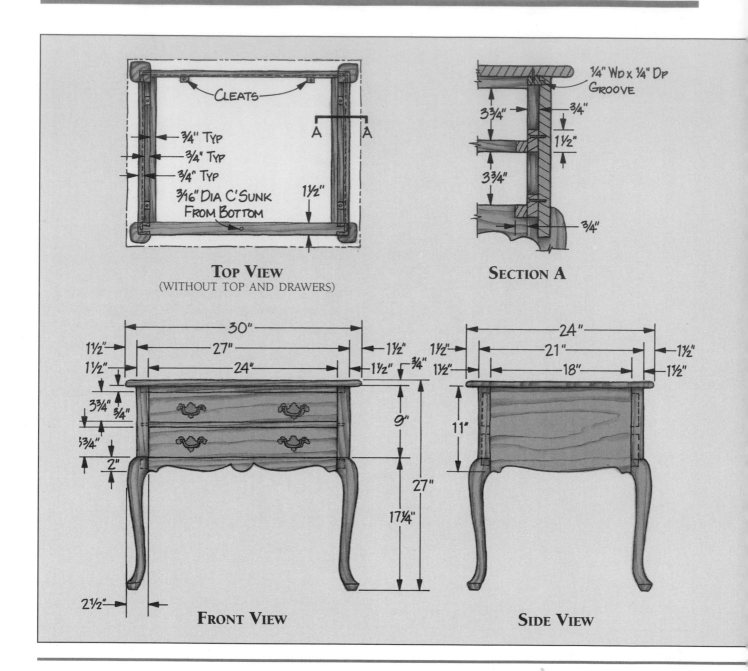

TOP VIEW
(WITHOUT TOP AND DRAWERS)

SECTION A

FRONT VIEW

SIDE VIEW

PLAN OF PROCEDURE

1 Select the stock and cut the parts to size.
To make this project, you need approximately 8 board feet of 12/4 (twelve-quarters) stock, 20 board feet of 4/4 (four-quarters) stock, and ¼ sheet (2 feet by 4 feet) of ¼-inch cabinet-grade plywood. Of the 4/4 stock, 14 board feet should be a cabinet-grade hardwood, while the remaining 6 board feet can be an inexpensive "secondary" wood. You can use almost any species of wood to build this project, but American Queen Anne furniture is traditionally made from

mahogany, walnut, or cherry. The side table shown is made from walnut, with maple as the secondary wood.

Plane the 12/4 stock to 2½ inches thick and rip four pieces for the legs, each 2½ inches wide. Cut the legs to length. Rip a 1½-inch-wide piece from the remaining 2½-inch-thick stock to make the bottom rail, and cut it to size.

Plane the 4/4 cabinet-grade stock to ¾ inch thick and cut the top and middle rails, side aprons, and back aprons. Glue up a wide board for the top and

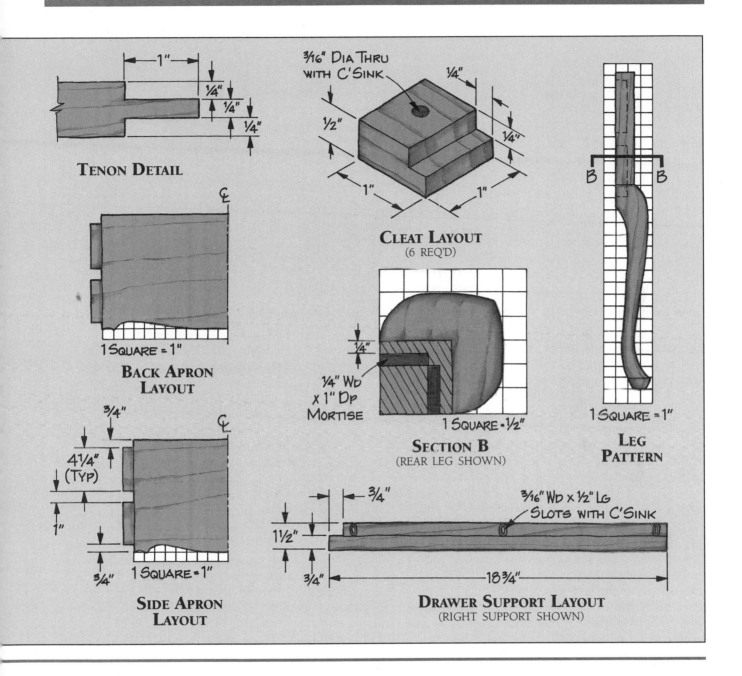

TENON DETAIL

BACK APRON LAYOUT
1 SQUARE = 1"

SIDE APRON LAYOUT
1 SQUARE = 1"

3/16" DIA THRU with C'SINK

CLEAT LAYOUT
(6 REQ'D)

1/4" WD x 1" DP MORTISE

SECTION B
(REAR LEG SHOWN)
1 SQUARE = 1/2"

LEG PATTERN
1 SQUARE = 1"

3/16" WD x 1/2" LG SLOTS WITH C'SINK

DRAWER SUPPORT LAYOUT
(RIGHT SUPPORT SHOWN)

set aside the remaining cabinet-grade stock for test pieces and drawer fronts. Plane the 4/4 secondary stock to ¾ inch thick and cut the drawer guides and drawer supports to size. Plane the remaining stock to ½ inch thick and set it aside to make the drawer sides, drawer backs, and cleats.

2 Cut the mortises in the legs. Lay out the posts of the cabriole legs on the leg stock, then mark the positions of the mortises. Remember, the back legs

are mortised differently than the front legs, since the mortise-and-tenon joints that connect the rails to the leg posts are different sizes than those that attach the aprons. Do *not* lay out the dovetail mortises for the top rail yet; wait until after you cut the dovetail tenons.

Cut the 1-inch-deep mortises in the legs. There are several ways to do this, but one of the easiest is to drill a line of overlapping ¼-inch-diameter holes for each mortise, then clean up the sides and corners with a chisel.

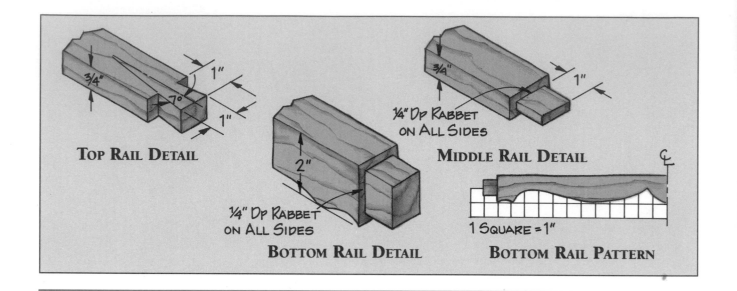

TOP RAIL DETAIL

¾" 7° 1" 1"

BOTTOM RAIL DETAIL

¼" DP RABBET ON ALL SIDES

2"

MIDDLE RAIL DETAIL

¾" 1"

¼" DP RABBET ON ALL SIDES

BOTTOM RAIL PATTERN

1 SQUARE = 1"

3 Cut the tenons in the rails and aprons. Lay out the tenons on the ends of the rails and aprons, as shown in the *Tenon Detail, Side Apron Layout, Back Apron Layout, Top Rail Detail, Middle Rail Detail,* and *Bottom Rail Detail.*

With the exception of the dovetail tenons on the top rail, all of the tenons can be formed by cutting two or more 1-inch-wide, ¼-inch-deep rabbets in the ends of the boards. Set up your table saw or table-mounted router to make these cuts, test the setup, then cut the tenons.

The tenons on the ends of the aprons must be "split" into two smaller tenons, as shown in the *Side Apron Layout* and *Back Apron Layout.* Two or more narrow tenons often hold better than a single wide one. Using the band saw, cut the notches required to split the tenons and create the top and bottom shoulders. While you're working at the band saw, cut the dovetail tenons in the top rail.

4 Cut the dovetail mortises. Using the top rail dovetail tenons as templates, mark the dovetail mortises on the top ends of the front legs. Remove as much waste as you can from the mortises with a drill and a Forstner bit, then cut away the rest with a chisel.

5 Cut the grooves in the aprons. The top is held to the aprons by L-shaped cleats. These cleats fit into grooves near the top edges of the aprons, as shown in *Section A,* and are screwed to the top. Cut the ¼-inch-wide, ¼-inch-deep apron grooves with a table saw or table-mounted router.

6 Make the cleats. Cut ¼-inch-wide, ¼-inch-deep rabbets in the ends of a ½-inch-thick board. These will create ¼-inch-thick tenons, as shown in the *Cleat Layout.* Saw the rabbeted edge off the stock, then cut up this piece to make six cleats.

7 Cut the cabriole shapes in the legs. Enlarge the *Leg Pattern* and trace it onto a scrap of thin plywood or hardboard. Cut the template on a band saw and sand the sawed edges. All the curves on the template should be smooth and "fair," with no flat spots or bumps.

Trace the cabriole shape onto the two *inside* surfaces of each leg. Then cut the shape on the band saw with a ¼-inch skip-tooth blade, using the compound cutting technique described in "Making Compound Cuts" on page 75.

8 Cut the shapes of the aprons and bottom rail. Enlarge the shapes of the aprons and bottom rail, shown in the *Side Apron Layout, Back Apron Layout,* and *Bottom Rail Pattern.* Trace these onto the stock and cut them with a band saw. Sand the sawed edges.

9 Round the edges of the tabletop. To soften the edges of the table, round-over the arrises with a router or shaper. You can create a simple round (as shown in the drawings) by making ⅜-inch-radius quarter-round cuts in the top and bottom arrises. Or, cut a thumbnail shape by making a ½-inch-radius quarter-round in the top arris and a ¼-inch-radius quarter-round in the bottom.

10 **Drill holes in the top rail and cleats, and slots in the drawer guides.** The top rail and cleats are screwed to the top. Drill ³/₁₆-inch-diameter pilot holes in these parts, then countersink the holes.

The drawer guides are also attached with screws, but these screws rest in *slots*. The slots allow you to raise or lower the drawer support assemblies, adjusting them so the drawers slide smoothly. Make the slots by drilling several overlapping holes, then countersink the slots. (*SEE FIGURE 8-1.*)

11 **Assemble the table.** Dry assemble the table to test the fit of the joints. When you're satisfied that the parts all fit properly, disassemble the table and finish sand all the outside surfaces. Round the knees of the cabriole legs with a disk sander, and shape the ankles with a drum sander. (*SEE FIGURES 8-2 AND 8-3.*)

Assemble the legs, aprons, and rails with glue. As you clamp up the assembly, check that the rails and aprons are square to one another. Also, glue the drawer supports to the drawer guides.

Let the glue dry, then attach the drawer guides to the inside surfaces of the side aprons with 1¹/₄-inch-long flathead screws, as shown in *Section A*. Center the top on the table and attach it to the aprons with cleats and 1-inch-long flathead screws. Using an off-set screwdriver, drive a 1¹/₄-inch-long flathead screw up through the top rail and into the top.

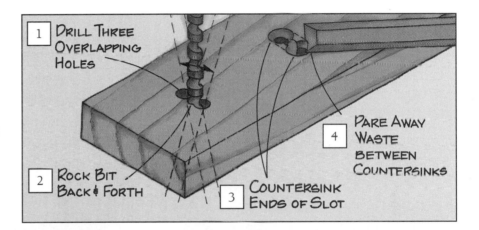

8-1 To make each slot in the drawer guides, drill three overlapping ³/₁₆-inch-diameter holes in a line. After drilling the last hole, work the drill bit back and forth to clean up the sides of the slot. To counter sink the slot, drill a countersink at each end, then join them by cutting a bevel between the countersinks with a chisel.

1 DRILL THREE OVERLAPPING HOLES

2 ROCK BIT BACK & FORTH

3 COUNTERSINK ENDS OF SLOT

4 PARE AWAY WASTE BETWEEN COUNTERSINKS

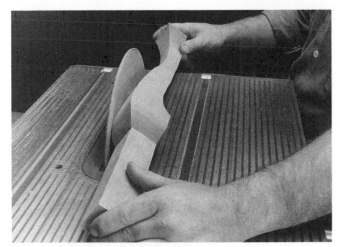

8-2 The knees of the cabriole legs are rounded to blend into the aprons and bottom rail. Do this rounding with a disk sander, being careful not to remove stock from any other part of the leg. If you don't have a disk sander, you can use a hand plane instead.

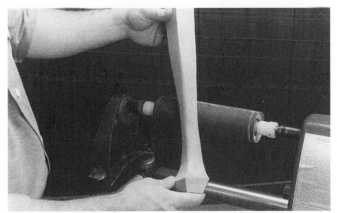

8-3 As the cabriole shape tapers down from the knees, the legs gradually become more round until they form a circle at the ankles. Then they flare out again to square feet. You can create this gradual rounding with an inflatable drum sander, as shown, or a spokeshave. Whatever tool you use, be careful to maintain the fair curves of the legs.

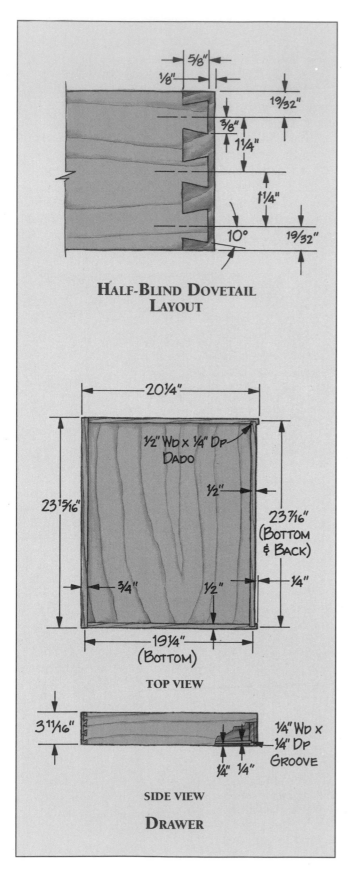

**HALF-BLIND DOVETAIL
LAYOUT**

5/8"
1/8"
19/32"
3/8"
1 1/4"
1 1/4"
10°
19/32"

20 1/4"
1/2" WD x 1/4" DP
DADO
1/2"
23 15/16"
23 7/16"
(BOTTOM
& BACK)
3/4"
1/2"
1/4"
19 1/4"
(BOTTOM)

TOP VIEW

3 11/16"
1/4" WD x
1/4" DP
GROOVE
1/4" 1/4"

SIDE VIEW

DRAWER

12 Make the drawers. Measure the drawer openings in the table. If the dimensions have changed from those shown in the working drawings, adjust the sizes of the drawer fronts, sides, backs, and bottoms.

Join the drawer sides to the drawer fronts with half-blind dovetails, as shown in the *Half-Blind Dovetail Layout.* If you have a router and a dovetail template, you can rout these joints. If not, cut the tails in the sides on the band saw. Use the tails as a template to lay out the pins in the fronts. Form these pins in the same manner that you made the dovetail mortises in the legs — remove as much waste as you can from between the pins with a drill and a Forstner bit, then cut away the rest with a chisel.

Cut 1/2-inch-wide, 1/4-inch-deep dadoes in the sides to hold the drawer backs. Also make 1/4-inch-wide, 1/4-inch-deep grooves in the inside surfaces of the drawer fronts, backs, and sides to hold the drawer bottoms, as shown in the *Drawer/Top View* and *Drawer/Side View.*

Finish sand the inside and outside surfaces of the drawer parts. Assemble the fronts, backs, and sides with glue, but allow the bottoms to "float" in their grooves. Make sure the parts are all square to one another as you clamp them together.

FOR BEST RESULTS

To ensure that the assembled drawers will not be warped or twisted, clamp them to a perfectly flat surface as the glue dries.

13 Fit the drawers. After the glue dries, attach drawer pulls to the drawer fronts. Fit the drawers in their openings and slide them in and out. If the drawers won't fit or if they bind as they slide, sand or plane a little stock from the surfaces that are rubbing. You can also adjust the level of the drawer supports so the drawers will slide more easily.

14 Apply a finish to the completed table.
Remove the drawers from the table and the drawer pulls from the drawers. Also detach the top from the table. Do any necessary last-minute touch-up sanding, then apply a finish to all wooden surfaces except the drawer sides, bottoms, and backs. Be sure to finish the inside surfaces of the table and the underside of the top — this will help prevent the broad boards from cupping or warping. When the finish dries, reassemble the table.

9

Nesting Boxes

Of the many techniques for making band-sawed boxes, one of the simplest is to resaw slabs off a block of wood, then glue them back together. If you cut the slabs in the proper order, they will go back together so that you can barely tell you've made any cuts at all. The box will look like a solid block.

There is another advantage to this simple technique. Once you cut all the slabs you need to make one box, you still have a block of wood left over. If that block is big enough, you can use it to make a second box — and the second box will fit inside the first. In fact, you can make three or more boxes from a medium-size block of wood, and they will all nest inside one another.

Note: There is no materials list for this project, since the dimensions of the box parts depend on the size of the wood block you use.

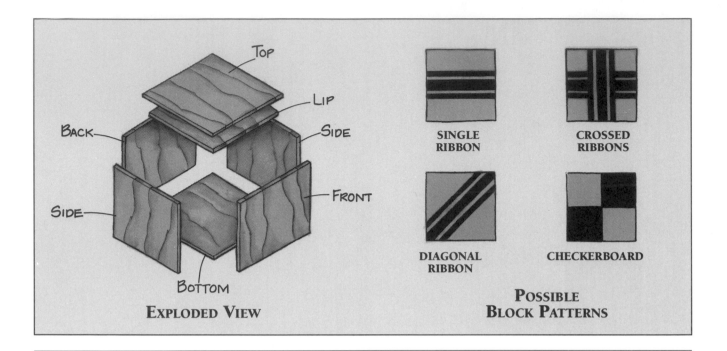

EXPLODED VIEW

POSSIBLE BLOCK PATTERNS

PLAN OF PROCEDURE

1 Glue up the block. Start with a wood block of almost any size. The only restriction is that you must be able to safely resaw the block on the band saw.

If you wish, add visual interest to this project by gluing up wood scraps of contrasting colors to make the block. Try to arrange them in a pattern that will remain roughly the same as the size of the block diminishes. The block that was used to make the boxes shown was designed to look like it had a ribbon tied around it. (*SEE FIGURES 9-1 AND 9-2.*) Each of the boxes cut from it looks like a gift-wrapped present. Several other designs are shown in the *Possible Block Patterns*.

2 Cut the parts of the boxes. Let the glue dry *at least* 24 hours before resawing the block. Then cut away thin slabs on the top, bottom, and sides to true up the glue joints. All six surfaces of the block must be perfectly flat and reasonably square to one another.

Decide how thick you want to make the parts of the box. The thinner you make them, the more boxes you can cut from a single block. On the other hand, if you make the parts too thin, the boxes won't be very strong. For adequate strength, the parts should be at least ¼ inch thick.

Set up the band saw to resaw slabs of the thickness you've chosen. With the block resting on its side, cut the top and bottom of the box from the block. Turn the block so it's resting on its bottom and resaw the front, back, and sides. Finally, turn it to rest on its side again and cut the lip. (*SEE FIGURE 9-3.*) Repeat this procedure to make parts for additional boxes.

3 Assemble the boxes. Sand the sawed surfaces of the box parts to remove the saw marks, then finish sand the *inside* surfaces. Glue the fronts, backs, and sides together, carefully matching the grain. Let the glue dry, then glue the bottoms to the assemblies. Don't worry if the bottoms seem too big for the boxes; this is due to the stock lost in the kerfs. Just match the grain as closely as possible. Later, you can sand off the protruding edges of the bottoms.

Also, glue the lips to the tops. Together, these two parts form the lids. Again, match the wood grain. When the parts are completely assembled, the lids should neatly fit the boxes. In addition, the smaller boxes should nest inside the larger ones.

4 Finish the boxes. Place the lids on the boxes and sand protruding edges of the top and bottom flush with the front, back, and sides. Also, sand all the glue joints clean and flush. Finish sand the outside surfaces and do any necessary touch-up sanding to the inside surfaces. Then apply several coats of finish to *all* surfaces, inside and out.

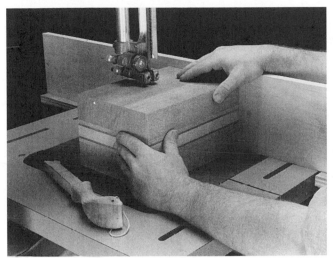

9-1 To make the crossed ribbon pattern shown, first glue up a single ribbon using slabs of contrasting woods. Let the glue dry completely, then cut the block in half *perpendicular* to the ribbons.

9-2 True up the cut surfaces of the block with a disk sander, then glue the blocks of wood back together with several additional slabs of contrasting wood between them. Some of the wood grains will be opposed, and this may create stress when the block expands and contracts. However, you can minimize the stress by slicing the parts of the boxes fairly thin, then sealing them inside and out with a finish.

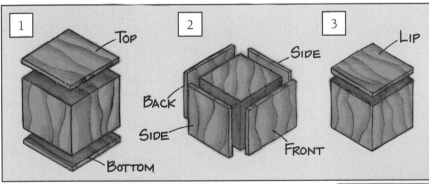

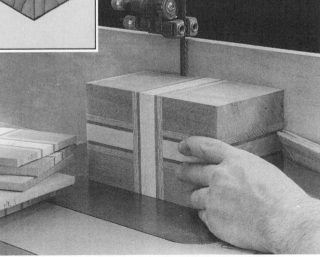

9-3 Resaw the box parts from the block in three steps. First, cut the top and the bottom with the block resting on its side. Second, cut the front, back, and sides with the block resting on its bottom. Finally, cut the lip with the block resting on its side again. Carefully mark the parts as they come off the block so you know what they are.

10

COMPOUND-CUT CHESS SET

There are thousands of designs for chess pieces, most of them either carved or turned on a lathe. But here's something a little different — a chess set that's sculpted entirely on the band saw.

Each piece is made using a special compound cutting technique.

The storage case and chessboard are also made with the aid of several band saw techniques. The dark and light squares of the playing field are resawn from long strips of wood, and the corners of the case are joined with band-sawed finger joints. Even the case sides and the strips that line the inside are made in part with the band saw.

114

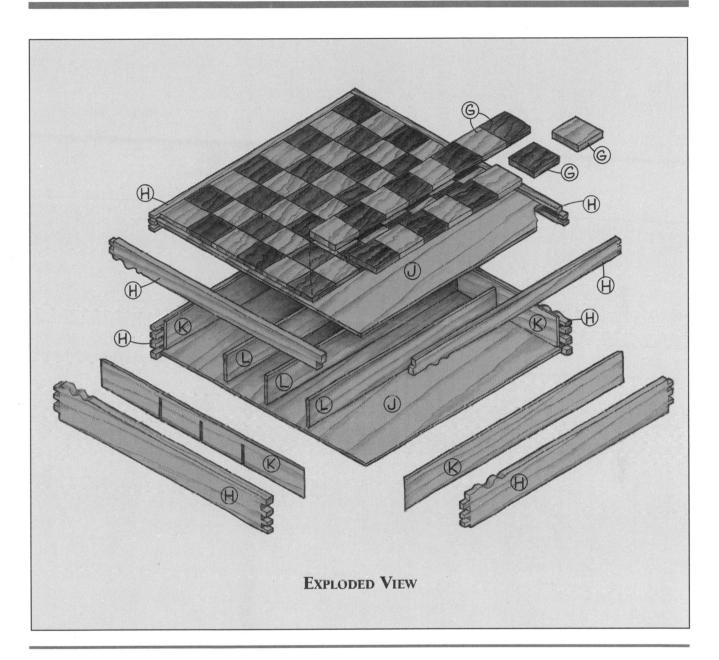

EXPLODED VIEW

MATERIALS LIST (FINISHED DIMENSIONS)

Parts

Chess pieces
A. Kings (2)   1³⁄₄″ x 1³⁄₄″ x 4¹⁄₄″
B. Queens (2)   1³⁄₄″ x 1³⁄₄″ x 4″
C. Bishops (4)   1¹⁄₂″ x 1¹⁄₂″ x 3¹⁄₂″
D. Knights (4)   1¹⁄₂″ x 1¹⁄₂″ x 3″
E. Rooks (4)   1¹⁄₂″ x 1¹⁄₂″ x 3³⁄₈″
F. Pawns (16)   1¹⁄₄″ x 1¹⁄₄″ x 2⁵⁄₈″

Storage case
G. Playing field
    squares (64)   ¹⁄₈″ x 2″ x 2″
H. Sides (4)   ¹⁄₂″ x 3¹⁄₄″ x 17″
J. Top/bottom*
    (2)   ¹⁄₄″ x 16¹⁄₂″ x 16¹⁄₂″
K. Lining
    strips (4)   ¹⁄₈″ x 2¹⁄₄″ x 16″
L. Dividers (3)   ¹⁄₄″ x 1¹⁄₂″ x 16″

Make these parts from plywood.

Hardware

Leather or felt
   (approximately 4 square feet)

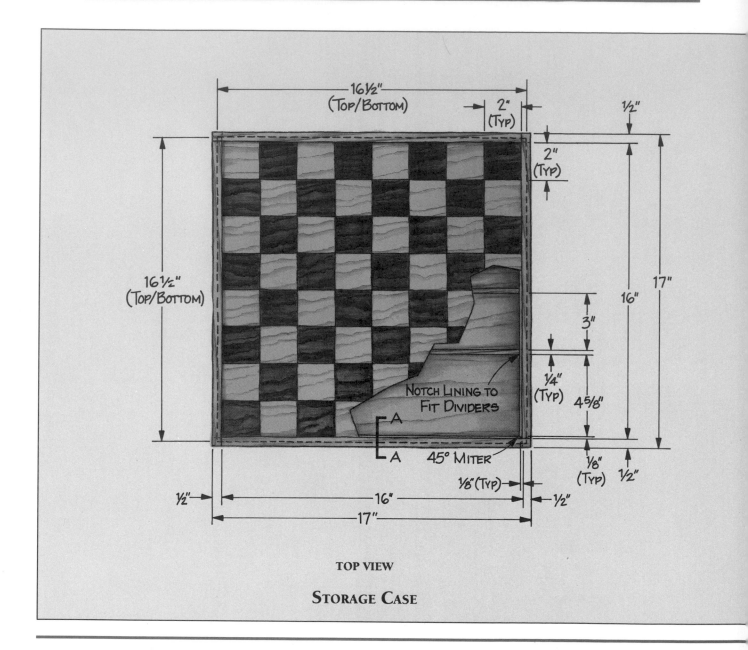

TOP VIEW

STORAGE CASE

PLAN OF PROCEDURE

1 Select the stock and cut the parts to size. To make the chess pieces and the playing field, you need about 4 board feet of 8/4 (eight-quarters) stock. This stock should be evenly divided between light and dark hardwoods. The set shown was made from 2 board feet of walnut and an equal amount of maple. When you select the stock, try to find boards that are a "strong" 8/4, since the playing field parts must be a full 2 inches thick after you plane them to size. If you can't find 8/4 stock that will do, purchase 10/4 (ten-quarters) stock.

To make the storage case on which the playing field is mounted, you need about 4 board feet of 4/4 (four-quarters) stock and ¼ sheet (2 feet by 4 feet) of ¼-inch cabinet-grade plywood. The case shown is made from curly cherry stock and birch plywood.

Plane the 8/4 stock to 2 inches thick and cut eight pieces 4 inches *wide* and 2⅛ inches *long*. (The grain must run in the 2⅛-inch direction.) Four pieces should be light wood; the other four should be dark. Set these aside to make the playing field squares.

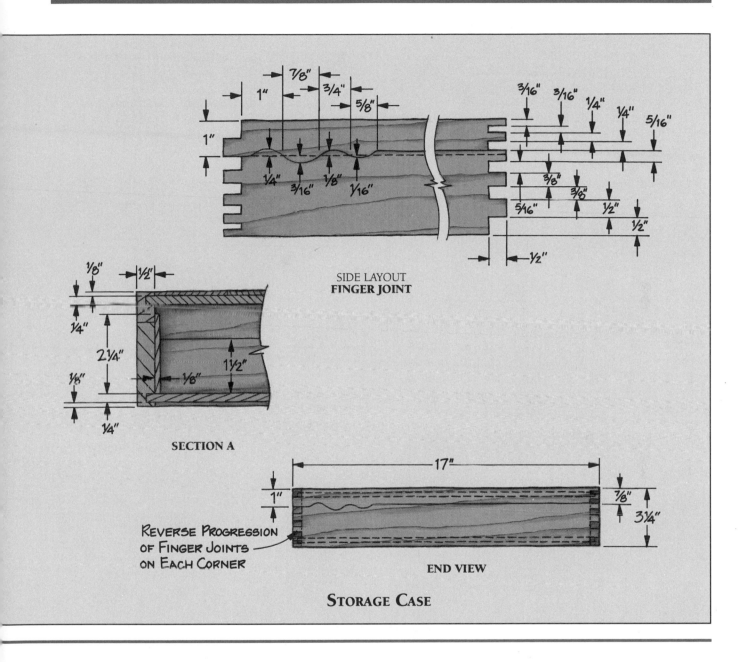

SIDE LAYOUT
FINGER JOINT

SECTION A

REVERSE PROGRESSION
OF FINGER JOINTS
ON EACH CORNER

END VIEW

STORAGE CASE

Set aside a small amount of the 2-inch-thick stock to use as test pieces or in case you make a mistake. Plane the remaining stock to 1¾ inches thick and cut the king and queen blanks. Plane the stock again, this time to 1½ inches thick, and cut blanks for the bishops, knights, and rooks. Finally, plane the stock to 1¼ inches thick and cut the pawn blanks. Remember, you must cut an equal number of light and dark blanks — one light and one dark queen, two light and two dark bishops, eight light and eight dark pawns, and so on.

Cut a 2½-inch-wide, 17-inch-long strip from the 4/4 stock and resaw it into four pieces of equal thickness. Plane the pieces to ⅛ inch thick and set them aside to make the lining strips. Cut two 2-inch-wide, 17-inch-long strips and resaw these in half; then plane these pieces to ¼ inch thick and set them aside to make the dividers.

Plane the remaining 4/4 stock to ½ inch thick and cut the sides to size. Also cut the top and bottom from ¼-inch plywood.

2 **Mark the shapes of the chess pieces on the stock.** Before you can cut the chess pieces, you must first make several *templates* to mark the shapes. Enlarge the patterns for the *King, Queen, Bishop, Knight, Rook,* and *Pawn.* Trace these onto thin scraps of plywood or hardboard and cut them out on a band saw with a 1/8-inch standard-tooth blade. Sand the sawed edges to remove any saw marks and to make the curves "fair" — that is, to smooth them out so there are no bumps or flat spots.

TRY THIS TRICK

To cut a small shape precisely, cut slightly wide of the line, then sand up to it.

Using the templates, trace the shape of each chess piece onto *two adjoining surfaces* of each blank, marking a face and a side. Be careful to center the pattern on each surface. For most pieces — the kings, queens, bishops, rooks, and pawns — mark the same shape

on both surfaces. The knights, however, require two different shapes. Mark the front shape on one surface and the side shape on another.

3 **Cut portions of the shapes on the table saw.** Although all the shapes of the chess pieces must be completed on the band saw, it's easier to cut some of the flat areas on the table saw *before* cutting the curves. These include the bevel at the bottom of all the chess pieces; the flat faces on the heads of the kings, queens, and rooks; and the notches at the top of the bishops and rooks.

FOR BEST RESULTS

Use a ripping blade with a raker grind to cut the notches in the heads of the bishops and rooks. These blades leave a flat-bottom kerf.

Before cutting good stock, make a test piece for each of the shapes you want to cut on the table saw. Set the blade height, blade angle, and fence position as closely as possible, then cut the test piece. If the cut isn't quite right, readjust the machine and cut the test piece again. Don't cut the good stock until you get the setup just right. (*SEE FIGURES 10-1 AND 10-2.*)

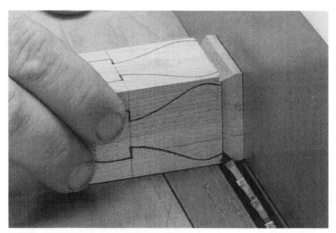

10-1 **To cut the bevels at the** bottoms of the chess pieces, set the table saw blade angle to 30 degrees. Adjust the blade height to cut just 1/8 inch deep and position the fence about 5/32 inch away from the blade. Cut a test piece to check the setup, sawing the bevel with the piece lying on its side and the bottom end against the fence. Use a large scrap of wood to help push the stock over the blade. When you're satisfied that the machine is set up properly, cut the bevel in all four sides of all 32 blanks.

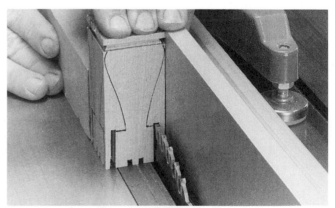

10-2 **You can also use the table** saw to cut the notches and flat surfaces on the heads of the kings, queens, bishops, and rooks. To make these cuts, rest the top end of each piece on the worktable and hold a side or a face against the fence. Use a large wood scrap to push the stock over the blade. Once again, check your setups with test pieces before cutting good stock.

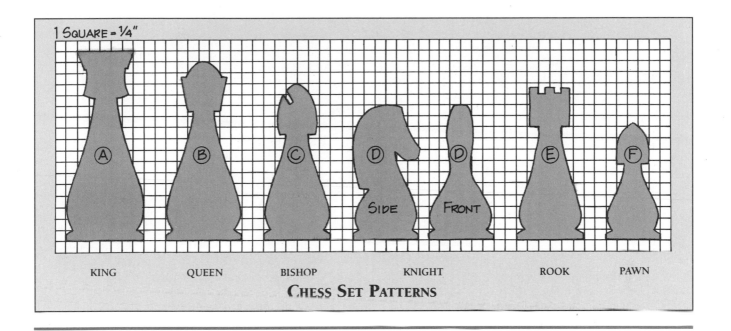

1 Square = ¼"

A — KING
B — QUEEN
C — BISHOP
D — KNIGHT (Side / Front)
E — ROOK
F — PAWN

CHESS SET PATTERNS

4 Complete the shapes. Cut the remaining portions of the chessmen shapes with the band saw and a ⅛-inch standard-tooth blade. To do this, use a variation on the compound cutting technique. As outlined in "Making Compound Cuts" on page 75, there are three basic steps to this technique:

1. Cut a pattern in the first surface.
2. Tape the waste back to the stock.
3. Turn the wood 90 degrees and cut a pattern in the second surface.

On this project, instead of taping the waste back to the stock, leave it attached with small *bridges* — tiny portions of the pattern that you don't cut until the very end. (SEE FIGURES 10-3 THROUGH 10-5.) With this change, the steps are slightly different:

1. Cut *most* of the pattern in the first surface, but leave small bridges to hold the waste in place.
2. Turn the wood 90 degrees and cut *most* of the pattern in the second surface. Once again, leave small bridges uncut.
3. Mentally, divide the pattern lengthwise into right and left halves. Cut the bridges on the *right-hand* half of one surface. Turn the stock 90 degrees to expose a new surface and cut the right-hand bridges *only.* Repeat for the third and fourth surfaces. To cut the last remaining bridges on the fourth surface, you may have to rest the stock on the waste blocks that fell away when you cut the third surface. This will hold the stock at the proper angle to the blade.

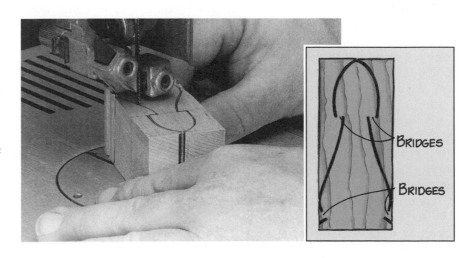

10-3 **When making compound** cuts to create the shapes of the chess pieces, leave *bridges* to hold the waste to the stock until you're ready to remove it. For example, when making a pawn, cut the shapes of the head, neck, and body in one surface, but don't cut the "chin" — the two flat surfaces under the head — or the spots just above the bevel. Leave these as bridges to hold the waste to the stock.

BRIDGES

BRIDGES

10-4 Turn the pawn 90 degrees
and cut the shapes in the second surface. Once again, leave bridges under
the chin and above the bevel.

10-5 When you've sawed as much
of the pawn shape as you can without removing the waste, mentally
divide the piece lengthwise into
right and left halves. Cut the two
bridges on one half. Turn the pawn
90 degrees and cut two more bridges.
Repeat until you've removed all the
waste.

By using bridges, you can cut all the chess pieces
without having to keep track of the waste or tape it
back to the stock. This saves a great deal of time (and
tape). **Note:** The bridging technique works well for
small or complex patterns — shapes in which there
are many short cuts. You can saw all but a small
portion at the end of these cuts, then easily back the
band saw blade out of the kerf. It's not efficient when
cutting large, simple patterns such as cabriole legs.

5 Sand the chess pieces. Sand the sawed surfaces
to remove the saw marks, then finish sand all surfaces.
This is no easy task when doing small pieces with
complex surfaces, but there are a few techniques you
can use to speed the process:

■ Use a strip sander to remove as many of the saw
marks as possible. Cut the belt so it's just 1/4 to 3/8 inch
wide. This will help you sand both the convex and
concave curves.

■ For tiny surfaces, such as the bevels and the
chins, smooth the wood with a sharp single-cut mill
file. Clean the file often with a file card to keep the
shavings from building up on it.

■ You can do a lot of the finish sanding by clamping a pad sander upside down in a vise, then hold-

ing the chess pieces against the vibrating pad. (*See
Figure 10-6.*)

■ If you don't care to preserve the crisp corners on
the chess pieces, use a flap sander to finish sand the
shapes. (*See Figure 10-7.*)

6 Make the playing field. Glue the eight pieces
of stock together that you had set aside to make the
playing field, alternating the light and dark woods.
(*See Figure 10-8.*) When looking at the glued-up block
from the edge, the strips of color should be 2 inches
side to side, but 2 1/8 inches top to bottom. To true up
the glue joints, resaw the block on a band saw and
sand the sawed surfaces on a disk sander. When you've
finished, the strips of color should be 2 inches square
and both the top and bottom surfaces of the block
should be perfectly flat.

Joint one side of the block. With the block resting
on the bottom surface and the jointed side against the
fence, resaw a strip 5/32 inch thick, 2 inches wide, and
16 inches long. Joint the sawed side again and resaw
another strip. Repeat this process until you have made
eight strips. Do *not* glue the strips together yet.

7 Join the corners of the case. The sides of the case are joined at the corners with "progressive" finger joints. The fingers and notches are very narrow near one edge of the board, then become progressively wider toward the other edge. The progression flops at each corner — the narrow fingers are near the top edge at one corner, then near the bottom at the other.

Lay out the finger joints on the ends of the sides, as shown in the *Finger Joint/Side Layout*. Then cut the finger joints on the band saw, as described in "Band Saw Joinery" on page 84.

10-6 To finish sand the surfaces of the chess pieces, clamp a pad sander upside down in a vise. Hold the pieces against the vibrating pad as you turn them this way and that. The pad sander won't reach all the surfaces — you'll still need to do some hand sanding — but it will smooth most of them.

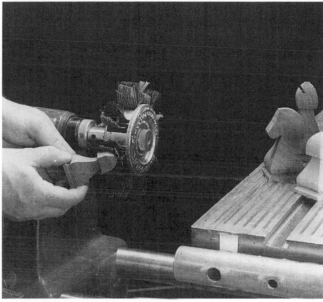

10-7 You can also use a flap sanding attachment to finish sand the chess pieces. The fingers of a flap sander will reach almost all the surfaces, but they will also round the corners. If you want to preserve the crisp corners on the pieces, don't use this tool.

10-8 When gluing up the stock for the playing field, the orientation of the wood strips is extremely important. Alternate the light and dark strips and join them so the end grain is at the top and bottom of the block. After the glue dries, remove a little stock from both the top and bottom so each strip will be perfectly square when viewed from the side.

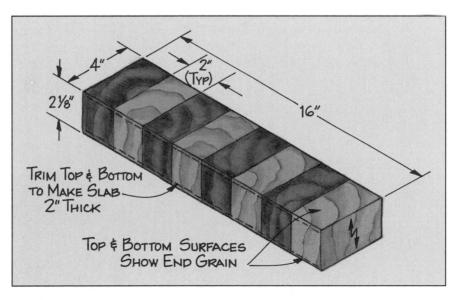

4"
2" (TYP)
2⅛"
16"
TRIM TOP & BOTTOM TO MAKE SLAB 2" THICK
TOP & BOTTOM SURFACES SHOW END GRAIN

8 **Rout the grooves in the side to hold the top and bottom.** The top and bottom of the storage case rest in ¼-inch-wide, ¼-inch-deep grooves near the edges of the sides. These grooves are *blind* — they aren't cut all the way through to the ends. Make these grooves with a table-mounted router, halting just before you cut through to the ends.

9 **Separate the sides into the top and bottom sections.** After you cut the joinery in the sides, rip the sides into the top and bottom sections. Lay out the cuts in the sides, as shown in the *Finger Joint/Side Layout.* The little wiggle in the cut is not necessary, but it adds visual interest to the case. Make the cuts on the band saw with a ⅛-inch standard-tooth blade — by leaving a narrow kerf, this blade will not diminish the size of the fingers noticeably.

10 **Assemble the bottom section.** Test fit the parts that you have made so far — sides, top, bottom, and playing field strips. The top surfaces of the playing field strips should protrude slightly above the top edges of the sides, and all joints should fit snugly. Hand fit any joints that need work, then finish sand the sides and bottom. **Note:** The jointed surfaces of the playing field strips should face *down.* Later, you'll plane and sand the sawed surfaces.

Glue the bottom portions of the sides and the bottom together. As you clamp up the assembly, check that the sides are square to one another.

11 **Install the lining strips and the dividers in the bottom section.** Carefully measure the inside dimensions of the bottom assembly. If the dimensions have changed slightly from those shown in the working drawings, adjust the sizes of the lining strips and dividers to compensate. Cut these parts to size, mitering the ends of the strips. Cut ¼-inch-wide, 1½-inch-long notches in the bottom edges of two opposing lining strips to fit over the dividers. Finish sand the lining strips and the dividers, then glue them in place.

12 **Assemble the top section.** Let the glue dry on the bottom assembly, then drape waxed paper over the corners. Glue and clamp the top portions of the sides, the top, and the playing field strips together on the bottom assembly. (*SEE FIGURE 10-9.*) This will ensure that the assembled top fits the bottom precisely. The waxed paper will prevent the top and bottom assemblies from sticking together.

13 **Finish the chess pieces and storage case.** Hand plane the playing field squares until they are flush with the top edge of the sides. Then finish sand the outside surfaces of the storage case and do any necessary touch-up sanding to the inside surfaces.

Apply a finish to both the storage case and the chess pieces. It's not necessary that you use the same finish on both the case and the pieces. On the set shown, the case is finished with five coats of tung oil (inside *and* outside) to help control the expansion and contraction of the playing field squares. The pieces, however, are finished with a brushing lacquer. The brushing lacquer is much easier to rub out on the small, contoured surfaces.

14 **Line the inside of the case and cover the bottoms of the pieces.** To help protect the finish on the chess pieces, apply leather or felt to the inside surfaces of the bottom and top. To protect the playing field, apply leather or felt to the bottom of the chess pieces. Use contact cement to adhere leather to wood, and polyvinyl (white) glue or rubber cement to adhere felt.

10-9 **Assemble the top section on** the bottom section to get a good fit. Use bar clamps to hold the sides together, as shown. Clamp long scraps of wood to the sides to hold the playing field strips against the top. Remember, the jointed surfaces of these strips should face down.

INDEX

Note: Page references in *italic* indicate photographs or illustrations.
Boldface references indicate charts or tables.

WOODWORKING GLOSSARY

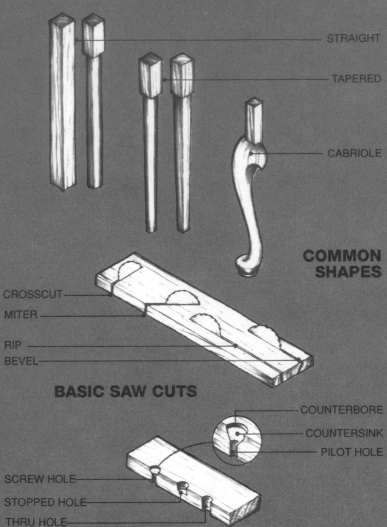

STRAIGHT

TAPERED

CABRIOLE

COMMON SHAPES

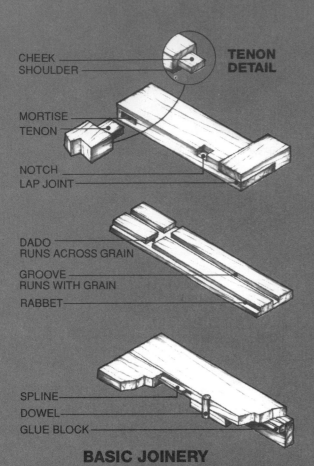

TENON DETAIL

CHEEK
SHOULDER

MORTISE
TENON

NOTCH
LAP JOINT

DADO
RUNS ACROSS GRAIN

GROOVE
RUNS WITH GRAIN

RABBET

SPLINE
DOWEL
GLUE BLOCK

BASIC JOINERY

CROSSCUT
MITER

RIP
BEVEL

BASIC SAW CUTS

COUNTERBORE
COUNTERSINK
PILOT HOLE

SCREW HOLE
STOPPED HOLE
THRU HOLE

HOLES

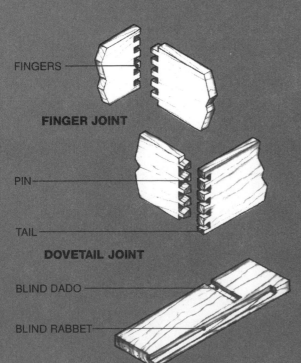

FINGERS

FINGER JOINT

PIN

TAIL

DOVETAIL JOINT

BLIND DADO

BLIND RABBET

SPECIAL JOINERY

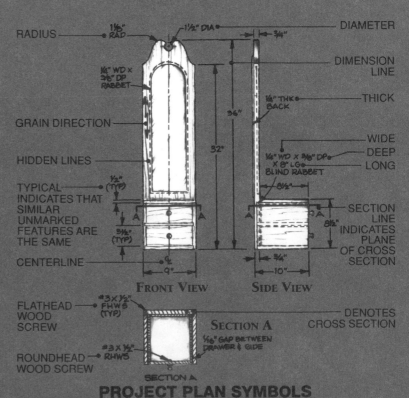

RADIUS — 1⅛" RAD — 1½" DIA — ¾" — DIAMETER

DIMENSION LINE

¼" WD x ⅜" DP RABBET

36"

⅛" THK BACK — THICK

32"

GRAIN DIRECTION

HIDDEN LINES

¼" WD X ⅜" DP X 8" LG BLIND RABBET

WIDE
DEEP
LONG

TYPICAL INDICATES THAT SIMILAR UNMARKED FEATURES ARE THE SAME

½" (TYP)

8½"

3½" (TYP)

SECTION LINE INDICATES PLANE OF CROSS SECTION

8½"

CENTERLINE

9"
9"

¾"
10"

FRONT VIEW   **SIDE VIEW**

FLATHEAD WOOD SCREW

#3 X ½" FHWS (TYP)

SECTION A

DENOTES CROSS SECTION

1/16" GAP BETWEEN DRAWER & SIDE

ROUNDHEAD WOOD SCREW

#3 X ½" RHWS

SECTION A

PROJECT PLAN SYMBOLS